LARSON, BOSWELL, KANOLD, STIFF

Passport
to Mathematics

BOOK 1

Bridge Unit
by Robyn Silbey

This Bridge Unit is intended to help students make the transition from elementary school mathematics to middle school mathematics. It contains brief lessons on important mathematical concepts and skills taught in elementary school and needed for *Passport to Mathematics, Book 1*. To support each Bridge Unit lesson, there are warm-up questions, class activities, and practice worksheets. Also included are a diagnostic test, an ongoing project, and assessments.

McDougal Littell
A HOUGHTON MIFFLIN COMPANY
Evanston, Illinois • Boston • Dallas

ISBN: 0-395-89642-8

23456789-PBO-02 01 00 99 98

Contents

Pacing and Planning

The pacing chart below is a summary of the days spent teaching the 12 chapters in *Passport to Mathematics, Book 1* as recommended in the planning guides found in the Teacher's Edition of that book. The pacing chart is based on a school year consisting of 160 days for a regular schedule and 80 days for a block schedule.

Chapter	1	2	3	4	5	6	7	8	9	10	11	12
Regular Schedule (days)	13	13	15	14	12	13	15	13	11	14	13	14
Block Schedule (days)	6.5	6.5	7.5	7	6	6.5	7.5	6.5	5.5	7	6.5	7

To allow for the time needed to teach lessons from this Bridge Unit, two suggested pacing charts are shown below. The first chart is based on teaching only two of the four topics in the Bridge Unit. (To determine which two topics will be taught, teachers should use the diagnostic test that is part of this Bridge Unit.) The 10 regular days and 5 block days needed to teach a partial Bridge Unit are made available by skipping textbook lessons 9.3, 11.4–11.6, 12.3, and 12.4, all of which cover mathematical topics not essential to a typical 6th grade curriculum.

Chapter	Bridge Unit	1	2	3	4	5	6	7	8	9	10	11	12
Regular Schedule (days)	10	13	13	15	14	12	13	15	13	10	14	8	10
Block Schedule (days)	5	6.5	6.5	7.5	7	6	6.5	7.5	6.5	5	7	4	5

Pacing and Planning

The pacing chart below allows time for teaching the complete Bridge Unit. The time needed to do this (20 regular days, 10 block days) is made available by skipping textbook lessons 9.3, 11.4–11.6, and 12.1–12.6. A planning guide for teaching the complete Bridge Unit appears on the pages that follow.

Chapter	Bridge Unit	1	2	3	4	5	6	7	8	9	10	11	12
Regular Schedule (days)	20	13	13	15	14	12	13	15	13	10	14	8	0
Block Schedule (days)	10	6.5	6.5	7.5	7	6	6.5	7.5	6.5	5	7	4	0

Pacing and Planning

Reg. Day	Block Day	Topic & Lesson	Materials	Suggested Daily Work
1	1	Topic 1 Lesson 1		**Project Introduction** **Warm-Up and Project** **Lesson 1: Rounding Numbers**
2		Topic 1 Lesson 1	newspapers, highlighter pens, scissors, tape, paper	**Activity 1** **Basic Practice 1**
3	2	Topic 1 Lesson 2	grocery store ads, 10 or 12 slips of paper of the same size, calculator (optional), paper	**Warm-Up and Project** **Lesson 2: Estimating Sums and Differences** **Activity 2** **Basic Practice 2**
4		Topic 1 Lesson 3	index cards, calculator (optional), paper	**Warm-Up and Project** **Lesson 3: Number Patterns in Multiplication and Division** **Basic Practice 3**
5	3	Topic 1 Lesson 4	four number cubes, a calculator	**Warm-Up and Project** **Lesson 4: Estimating Products** **Basic Practice 4**
6		Topic 1 Lesson 5	blank number grid (see blackline master), paper	**Warm-Up and Project** **Lesson 5: Estimating Quotients** **Activity 5** **Basic Practice 5**
7	4	Topic 2 Lesson 1		**Basic Assessment: Topic 1** **Warm-Up and Project** **Lesson 1: Divisibility** **Activity 1** **Basic Practice 1**

Pacing and Planning

Reg. Day	Block Day	Topic & Lesson	Materials	Suggested Daily Work
8		Topic 2 Lesson 2	grid paper (see blackline master)	**Warm-Up and Project** **Lesson 2: Greatest Common Factor** **Activity 2** **Basic Practice 2**
9	5	Topic 2 Lesson 3		**Warm-Up and Project** **Lesson 3: Prime and Composite Numbers** **Activity 3** **Basic Practice 3**
10		Topic 3 Lesson 1	paper, scissors	**Basic Assessment: Topic 2** **Warm-Up and Project** **Lesson 1: Fractions in Simplest Form** **Activity 1** **Basic Practice 1**
11	6	Topic 3 Lesson 2	10-by-10 grid paper (see blackline master)	**Warm-Up and Project** **Lesson 2: Changing Decimals to Fractions** **Activity 2** **Basic Practice 2**
12		Topic 3 Lesson 3	unlined paper, scissors	**Warm-Up and Project** **Lesson 3: Changing Fractions to Decimals** **Activity 3** **Basic Practice 3**
13	7	Topic 3 Lesson 4	a calculator (optional)	**Warm-Up and Project** **Lesson 4: Working with Values Greater Than 1** **Activity 4** **Basic Practice 4**
14		Topic 4 Lesson 1	scissors, unlined paper	**Basic Assessment: Topic 3** **Warm-Up and Project** **Lesson 1: Identifying Quadrilaterals** **Activity 1** **Basic Practice 1**

Reg. Day	Block Day	Topic & Lesson	Materials	Suggested Daily Work
15	8	Topic 4 Lesson 2	protractor, ruler	**Warm-Up and Project** **Lesson 2: Measuring and Constructing Angles** **Activity 2** **Basic Practice 2**
16		Topic 4 Lesson 3	a centimeter ruler, meter stick, paper, marker	**Project** **Lesson 3: Finding Perimeters of Polygons** **Activity 3** **Basic Practice 3**
17	9	Topic 4 Lesson 3	centimeter/millimeter ruler	**Warm-Up and Project** **Lesson 4: Finding Areas of Rectangles** **Activity 4** **Basic Practice 4**
18		Review		**Basic Assessment: Topic 4** **Project Summary**
19	10	Projects		**Project Presentations**
20		Assess		**Cumulative Assessment**

Bridge Unit, PASSPORT TO MATHEMATICS BOOK 1

Diagnostic Test

BRIDGE UNIT

OVER BRIDGE UNIT TOPICS

Topic 1: Estimation and Number Sense

Lesson 1

1. Round 56 to the nearest ten.

2. Round 791 to the nearest hundred.

3. Round 2407 to the nearest thousand.

4. Round 3.5 to the nearest whole number.

Lesson 2

Estimate each sum or difference by rounding each number to the place held by its first digit.

5. $305 + 679$

6. $0.8 + 2.3 + 5.1$

7. $9088 - 4572$

8. $62 - 16$

Lesson 3

Find each product or quotient.

9. a. 82×100 **b.** $82 \div 100$

10. a. 50×10 **b.** $50 \div 10$

Complete each statement.

11. 12 m $= \underline{\ ?\ }$ cm

12. 1500 m $= \underline{\ ?\ }$ km

Lesson 4

In Questions 13–16, use rounding to estimate each product. First round any number of two or more digits to the place held by its first digit.

13. 33×5

14. 67×42

15. 108×86

16. Mariko works 7 hours and 15 minutes five days a week. Estimate the total amount of time she works each week.

Lesson 5

Use rounding and compatible numbers to estimate each quotient.

17. $4725 \div 7$

18. $207 \div 3$

19. $636 \div 53$

20. $2356 \div 76$

Topic 2: Numbers and Their Factors

Lesson 1

Indicate by which of the digits 2, 3, 5, 9, and 10 each number is divisible.

21. 345

22. 882

23. 2760

24. 18,395

Lesson 2

25. Write all the factors of 42 from least to greatest.

26. Find the greatest common factor of 15 and 28.

27. Find the greatest common factor of 26 and 78.

Diagnostic Test

OVER BRIDGE UNIT TOPICS

28. Find the greatest common factor of 18, 48, and 60.

Lesson 3

Indicate whether the number is *prime* or *composite*.

29. 87   **30.** 91

Give the prime factorization of each number.

31. 56   **32.** 60

Topic 3: Fraction and Decimal Concepts

Lesson 1

Indicate whether each fraction is in simplest form. Write *yes* or *no*. If your answer is *no*, give the simplest form.

33. $\frac{20}{64}$ **34.** $\frac{77}{100}$ **35.** $\frac{27}{45}$ **36.** $\frac{40}{63}$

Lesson 2

Write an equivalent fraction for the decimal. Then write the fraction in simplest form if it is not already in simplest form.

37. 0.4   **38.** 0.95   **39.** 0.73   **40.** 0.02

Lesson 3

41. Write $\frac{724}{1000}$ as a decimal.

Rewrite each fraction with a denominator of 10, 100, or 1000. Then write the decimal equivalent.

42. $\frac{4}{5}$ **43.** $\frac{7}{8}$ **44.** $\frac{41}{50}$

Lesson 4

45. Write $5\frac{3}{4}$ as an improper fraction and as a decimal.

46. Write $\frac{23}{3}$ as a mixed number.

Write an equivalent improper fraction in simplest form for each decimal.

47. 6.56   **48.** 1.6

Topic 4: Geometry

Lesson 1

49. a. Sketch a rhombus.   **b.** Is every rhombus also a parallelogram? Explain.

Diagnostic Test

OVER BRIDGE UNIT TOPICS

50. Is a rectangle *always, sometimes,* or *never* a square? Explain.

51. Is a square *always, sometimes,* or *never* a rectangle? Explain.

52. Is it possible for a trapezoid to have two pairs of parallel sides? If so, sketch an example. If not, explain why not.

Lesson 2

Use a protractor to measure each angle to the nearest 5°.

53. 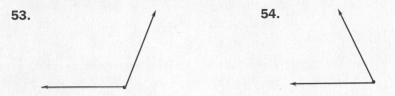   **54.**

55. Draw a triangle with one angle of 90° and another angle of 35°.

56. Predict the measure of the third angle of the triangle you drew in Question 55. Explain how you reached your answer. Then measure the angle to check your prediction.

Lesson 3

Find the perimeter of each polygon.

57. a square garden, sides of length 20 ft

58. a rectangle, length 14 m and width 10 m

Measure each side of the polygon to the nearest 0.5 cm. Then find the perimeter of the polygon.

59. 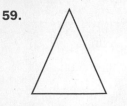   **60.**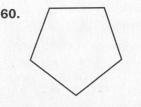

Lesson 4

In Questions 61–62, find the area of each rectangle or square.

61. a square garden, sides of length 20 ft   **62.** a rectangle, length 14 m and width 10 m

63. Find the area of the region shown.

64. The area of a rectangle is 32 square feet. Give two possible sets of dimensions for the rectangle.

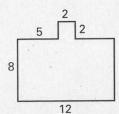

Bridge Unit, PASSPORT TO MATHEMATICS BOOK 1

3

1. 60
2. 800
3. 2000
4. 4
5. 1000
6. 8
7. 4000
8. 40
9. a. 8200
   b. 0.82
10. a. 500
    b. 5
11. 1200
12. 1.5
13. 150
14. 2800
15. 9000
16. 35 h (or 36 h)
17. 700
18. 70
19. 13
20. 30
21. 3, 5
22. 2, 3, 9
23. 2, 3, 5, 10
24. 5
25. 1, 2, 3, 6, 7, 14, 21, 42
26. 1
27. 26
28. 6
29. composite
30. prime
31. $56 = 2 \times 2 \times 2 \times 7$
32. $60 = 2 \times 2 \times 3 \times 5$
33. No; $\frac{5}{16}$.
34. Yes.
35. No; $\frac{3}{5}$.
36. Yes.
37. $\frac{4}{10} = \frac{2}{5}$
38. $\frac{95}{100} = \frac{19}{20}$
39. $\frac{73}{100}$
40. $\frac{2}{100} = \frac{1}{50}$
41. 0.724
42. $\frac{8}{10}$; 0.8
43. $\frac{875}{1000}$; 0.875
44. $\frac{82}{100}$; 0.82

45. $\frac{23}{4}$; 5.75
46. $7\frac{2}{3}$
47. $\frac{164}{25}$
48. $\frac{8}{5}$
49. a. Check students' work.
    b. Yes; every rhombus has two pairs of parallel sides.
50. Sometimes; a rectangle with all sides that are the same length is also a square. A rectangle whose length is unequal to its width is not a square.
51. Always; the angles of a square are all right angles and the opposite sides are parallel, so a square is also a rectangle.
52. No; a trapezoid has exactly one pair of parallel sides.
53. 110°
54. 65°
55. Sketches may vary. Sample:

56. 55°; the sum of the measures of the three angles is 180°, so the measure of the third angle is 180° − 90° − 35° = 55°. The prediction is correct.
57. 80 ft
58. 48 m
59. 2.5 cm, 2.5 cm, and 2 cm; 7 cm
60. All five sides are 1.5 cm long; 7.5 cm.
61. 400 square feet
62. 140 square meters
63. 100 square units
64. Sample answers: 8 ft by 4 ft, 16 ft by 2 ft, 32 ft by 1 ft

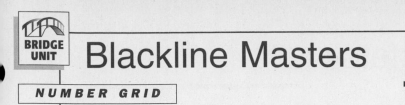

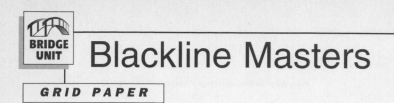
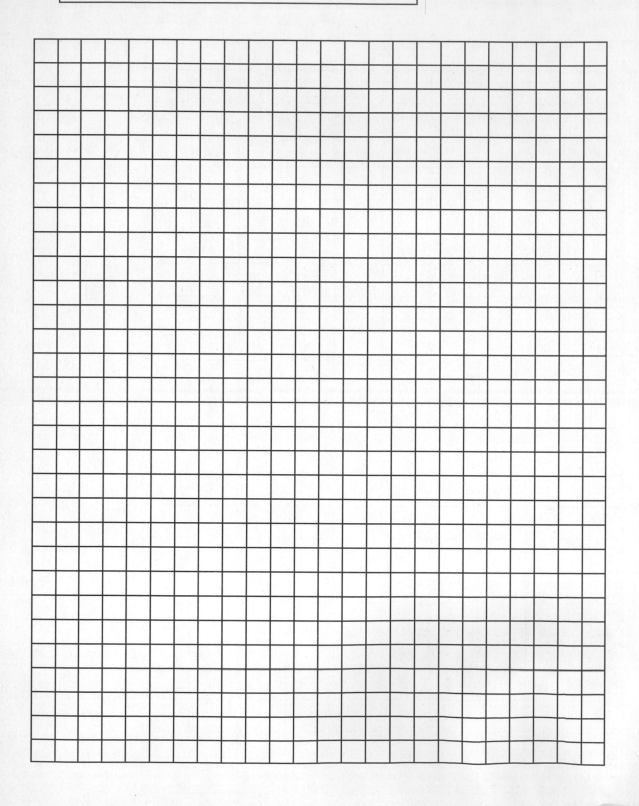

Blackline Masters

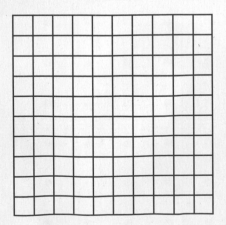

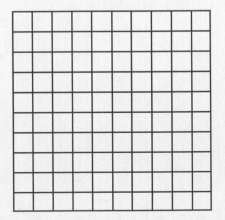

Blackline Masters

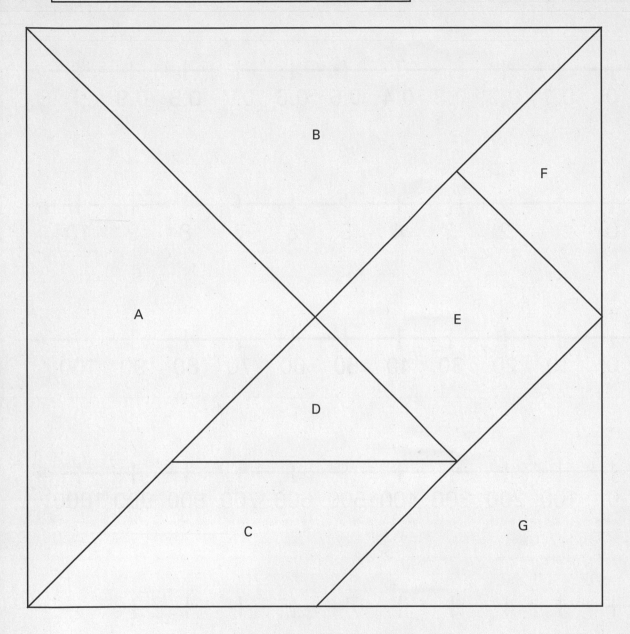

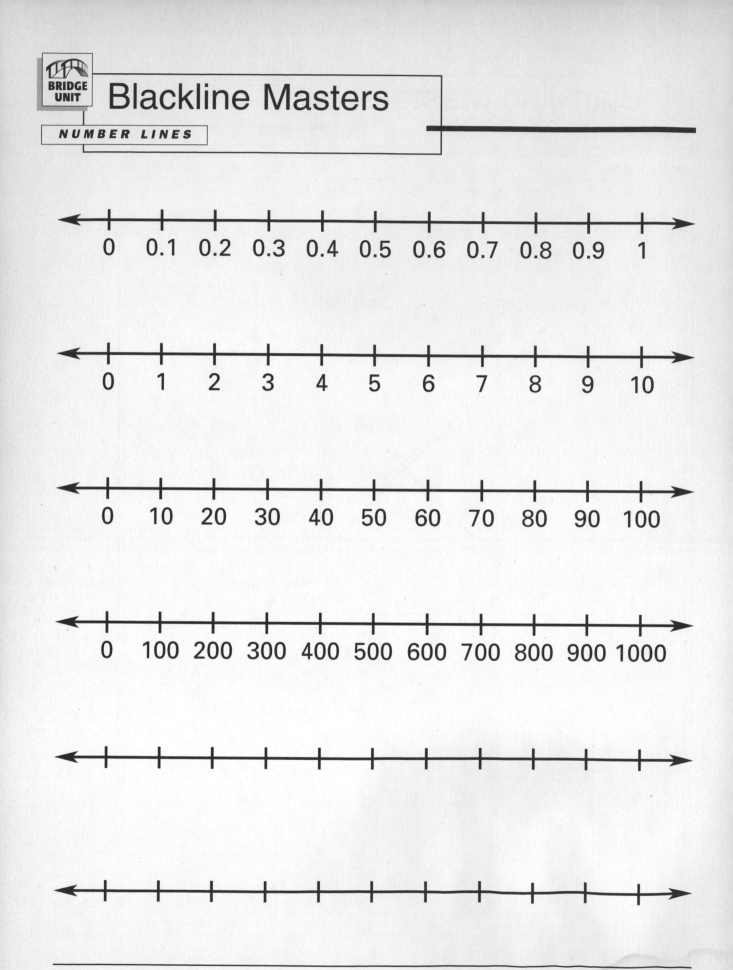

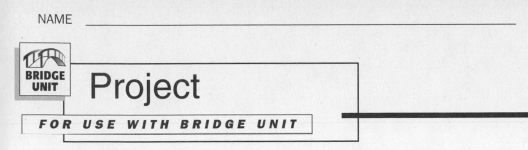

Project

FOR USE WITH BRIDGE UNIT

Design, order, and advertise for a school store

The merchant of a store needs to gather information about the customers who will shop at the store. Then the merchant can meet those customers' needs efficiently, affordably, and profitably. Your project is to plan a school store. You will design the layout of the store and make plans for ordering goods for the store. You will also create an advertisement that shares information with your customers—the students in your school.

Getting Started

For this project, you should work in a group of three or four students. First, you need to decide what your schoolmates may want to buy at a school store. You may want to sell supplies that students cannot purchase at an office supply store, or items that can be made specifically for students at your school. You will also need to think about what times of day students will have time to shop at the store. Decide what questions you should ask your schoolmates in order to determine the best chance of success for your school store.

Topic 1: Estimation and Number Sense

The five lessons covering this topic discuss how to make initial plans for your store. You will use the skills and concepts of these lessons throughout your project. You will gather information about costs of items that you may wish to sell in your school store. You will think about how you can price items so that your store makes a profit. You will also start thinking about the appearance of the school store and about scheduling workers for it.

Topic 2: Numbers and Their Factors

Now that you have thought about the items you plan to sell at the school store and the prices you hope to charge, you can begin planning the details. You will need to decide how to package each item to benefit the school store and the customers. As you complete the lessons covering this topic, you will also explore methods for decorating the school store bulletin board and arranging displays.

(continued)

Project *(continued)*

Topic 3: Fraction and Decimal Concepts

The lessons covering this topic enable you to become more specific about the details of managing a store. You will work with scheduling work times for student clerks and teacher supervisors. You will be introduced to surveys and other methods of gathering information about your store. This information can help you manage the store more effectively. You will also begin taking a look at keeping spare stock on hand to fill your product displays.

Topic 4: Geometry

You are now ready to finalize the physical layout of the school store. The geometry in these lessons will help you with the design and decoration of your store. You will decide how to arrange the store, and where to place tables and displays. You will work to make the store look inviting to customers and to fit goods to their display tables . Finally, you will decide what needs to be advertised to your customers and what data is simply needed for your records. Using this information, you will create an advertisement to attract customers to your new school store.

Goal
Round whole numbers to the nearest 10, 100, and 1000. Round decimal numbers to the nearest whole number.

Rounding Numbers

Shandra delivers 72 newspapers each morning before school. Shandra can use rounding to estimate the number of newspapers she delivers.

Terms to Know

Example / Illustration

Estimate a number close to an exact value that you use instead of the exact value, or, to use a number close to an exact value instead of the exact value	...60 61 62 63 64 65 66 67 68 69 70 71 72 73 74 75 76 77 78 79 80 81 82 83 84... Because 70 is close to 72, 70 is a good estimate for 72.
Rounding a method used to estimate a number or amount	To round 72 to the nearest 10, find the multiple of 10 that is closest to 72. 72 40 50 60 70 80 90 The closest multiple of 10 is 70. To the nearest 10, 72 rounds to 70.

UNDERSTANDING THE MAIN IDEAS

You can use rounding to estimate the size of a number.

(continued)

Example 1

The number of students at Rocky View Middle School is 683. Estimate the number of students at the school to the nearest hundred.

■ Solution ■

Step 1: Locate 683 between two multiples of 100.

The number 683 lies between 600 and 700.

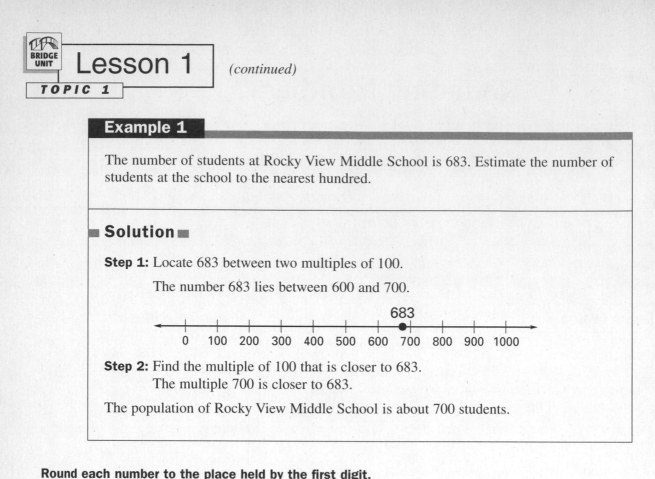

Step 2: Find the multiple of 100 that is closer to 683.
The multiple 700 is closer to 683.

The population of Rocky View Middle School is about 700 students.

Round each number to the place held by the first digit.

1. 28   **2.** 91   **3.** 567   **4.** 309   **5.** 1999   **6.** 6208

To round a decimal to the nearest whole number, locate the decimal between two whole numbers. Then decide which whole number is closer to the decimal.

Example 2

Harvey's meal contained 4.8 grams of fat. To the nearest whole number, about how many grams of fat were in Harvey's meal?

■ Solution ■

Step 1: Locate 4.8 between two whole numbers.

The number 4.8 lies between 4 and 5.

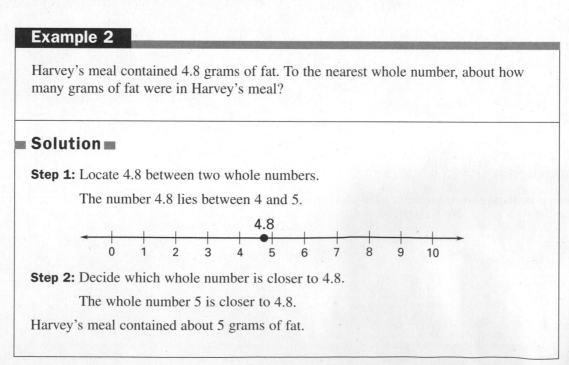

Step 2: Decide which whole number is closer to 4.8.

The whole number 5 is closer to 4.8.

Harvey's meal contained about 5 grams of fat.

(continued)

Bridge Unit, PASSPORT TO MATHEMATICS BOOK 1

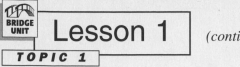
Round each decimal to the nearest whole number.

7. 2.3   **8.** 4.6   **9.** 8.1   **10.** 7.4

When a number is exactly in the middle of an estimate interval, round up to the greater number.

> Because 25 is exactly halfway between 20 and 30, round 25 *up* to 30.
> Because 650 is exactly halfway between 600 and 700, round 650 *up* to 700.
> Because 8.5 is exactly halfway between 8 and 9, round 8.5 *up* to 9.

Example 3

Jeremy sees a jacket he would like to buy that costs $85. What is the cost rounded to a multiple of $10?

■ Solution ■

Step 1: Locate the cost between two multiples of $10.

The cost is between $80 and $90.

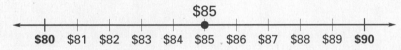

$85

$80 $81 $82 $83 $84 $85 $86 $87 $88 $89 $90

Step 2: Because the amount is exactly halfway between the two amounts, round up to the greater number.

Jeremy's jacket will cost about $90.

Round each number to the place held by the first digit.

11. 45   **12.** 95   **13.** 350   **14.** 750

15. 1500   **16.** 8500   **17.** 2.5   **18.** 5.5

....
Spiral Review

19. Write the number for $(3 \times 100) + (2 \times 10) + (6 \times 1)$.

20. Write three odd numbers between 4567 and 4575.

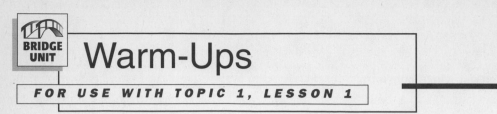

Warm-Ups

FOR USE WITH TOPIC 1, LESSON 1

Standardized Testing Warm-Ups

1. Which statement is true?

    A $2.5 < 2.1$ **B** $4.1 = 1.4$ **C** $6.0 > 5.8$ **D** $8.9 > 9.0$

2. Sally's family is displaying the word name for their house number above the front door. Their house number is 1051. Which word name will they display?

    A one hundred fifty-one

    B one thousand, fifteen

    C one thousand, fifty-one

    D one thousand, five hundred one

Homework Review Warm-Ups

3. Give the two closest multiples of 1000 and of 100 that the number 1361 lies between.

4. How far is 8.7 from 9?

5. Is 79.4 closer to 79 or to 80?

6. Is 200 or 300 the better estimate for 249? Why?

Project

FOR USE WITH TOPIC 1, LESSON 1

1. a. Survey your friends, classmates, and schoolmates to find out what they would buy from a school store.

    b. Organize your collected data so that you can decide which items you would like to purchase for the store.

    c. How do your organized data help you decide which items you should carry in your store? Could you make your decisions by estimating the results of the survey? Explain.

2. Use your conclusions to choose 10 items to sell in the store. Find the regularly priced cost for each item. Use newspapers, magazines, or advertisements to help you. Round the regularly priced cost of each item to the nearest ten cents or to the nearest dollar.

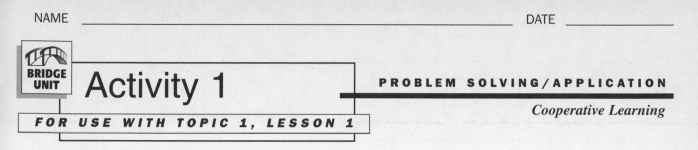

Activity 1

Numbers in the News

In this activity, you will study how reporters use estimated and exact numbers in the news.

You will need:

- several newspapers
- highlighter pens
- scissors
- tape
- plain paper

What to do:

Work with a partner. Look for newspaper articles containing both large and small numbers. You may wish to look through the front section, the business section, or the sports section. Cut out the article and tape it to a sheet of plain paper. Highlight each number in the article.

Write your answers on the sheet for each article.

1. Decide and record if you think the numbers in the article are estimates or exact amounts.

2. Look back at your answers to Exercise 1. How did you decide when a number was an estimate? In what ways are all the estimates alike?

3. How did you decide when a number was an exact amount?

4. Suppose you had written the article. Would you change any numbers from estimates to actual amounts? from actual amounts to estimates? Explain your answer.

5. In your opinion, when is it better to use estimates in news reporting? When is it important to use actual amounts?

Basic Practice 1

FOR USE WITH TOPIC 1, LESSON 1

Round each number to the nearest ten.

1. 43	**2.** 81	**3.** 37	**4.** 66	**5.** 95
6. 12	**7.** 38	**8.** 54	**9.** 39	**10.** 25

Round each number to the nearest hundred.

11. 275	**12.** 462	**13.** 586	**14.** 623	**15.** 250
16. 394	**17.** 826	**18.** 148	**19.** 909	**20.** 777

Round each number to the nearest thousand.

21. 4631	**22.** 2503	**23.** 7821	**24.** 8099	**25.** 2468
26. 8246	**27.** 6135	**28.** 6565	**29.** 4654	**30.** 1907

Round each decimal to the nearest whole number.

31. 2.1	**32.** 5.8	**33.** 6.7	**34.** 1.4	**35.** 8.2
36. 7.5	**37.** 3.3	**38.** 4.6	**39.** 2.9	**40.** 6.5

Write the least and greatest whole numbers that you can round to the given number.

41. 60, rounded to the nearest 10

42. 80, rounded to the nearest 10

43. 500, rounded to the nearest 100

44. 200, rounded to the nearest 100

45. 4000, rounded to the nearest 1000

46. 7000, rounded to the nearest 1000

In Exercises 47–51, write the least and greatest decimal numbers, to the tenths' place, that you can round to the given whole number.

47. 1	**48.** 6	**49.** 9	**50.** 10	**51.** 74

52. In your own words, write a step-by-step general method for rounding whole numbers. Then use the method you wrote to round 3468 to the nearest thousand.

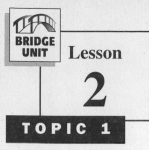

Estimating Sums and Differences

Goal
Use rounding to find estimated sums and differences. Solve problems by estimating sums and differences.

In Lesson 1, you learned how to use rounding to estimate numbers. You can use rounded numbers to estimate sums and differences.

Terms to Know

Terms to Know	Example / Illustration
Sum the answer in an addition problem	In the problem $7 + 3 = 10$, 10 is the sum.

Addends
    the numbers that are added together in an
    addition problem
	In the problem $9 + 11 = 20$,
	9 and 11 are the addends.

Difference the answer in a subtraction problem	In the problem $50 - 20 = 30$, 30 is the difference.

UNDERSTANDING THE MAIN IDEAS

You can use estimation to make sure your answer in an addition problem or a subtraction problem is reasonable. To estimate a sum, round the addends and use mental math to add.

(continued)

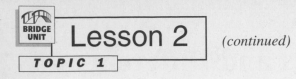
Example 1

A newspaper reported that about 1000 people attended a neighborhood flea market over a 3 day weekend. Was the report accurate?

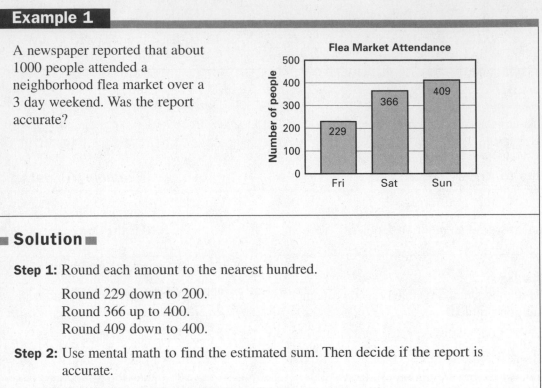

Flea Market Attendance

■ Solution ■

Step 1: Round each amount to the nearest hundred.

Round 229 down to 200.
Round 366 up to 400.
Round 409 down to 400.

Step 2: Use mental math to find the estimated sum. Then decide if the report is accurate.

$$200 + 400 + 400 = 1000$$

The newspaper report is accurate.

In Exercises 1–6, estimate each sum by first rounding each number to the place held by the first digit and then using mental math.

1. $57 + 31$
2. $19 + 42 + 68$
3. $825 + 352$
4. $287 + 391 + 405$
5. $1203 + 7915$
6. $4.4 + 5.1 + 1.9$

7. The flea market organizers expected about 500 people to attend on Friday and Saturday combined. Was their estimate correct? Explain your reasoning. Use the graph in Example 1 to help you.

(continued)

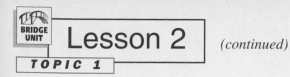

Lesson 2 *(continued)*

To estimate a difference, round the numbers and use mental math to subtract.

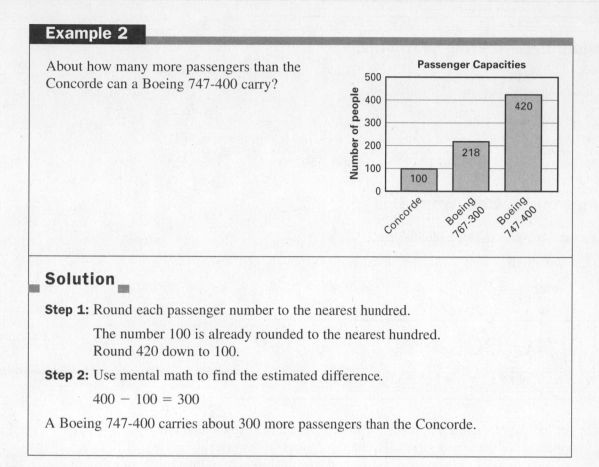

Example 2

About how many more passengers than the Concorde can a Boeing 747-400 carry?

Passenger Capacities

Concorde: 100, Boeing 767-300: 218, Boeing 747-400: 420

Solution

Step 1: Round each passenger number to the nearest hundred.

The number 100 is already rounded to the nearest hundred. Round 420 down to 100.

Step 2: Use mental math to find the estimated difference.

$$400 - 100 = 300$$

A Boeing 747-400 carries about 300 more passengers than the Concorde.

In Exercises 8–13, estimate each difference by rounding each number to the place held by the first digit and then using mental math.

8. $92 - 47$ **9.** $315 - 127$ **10.** $589 - 222$

11. $5089 - 1992$ **12.** $8.2 - 2.8$ **13.** $5.1 - 4.8$

14. About how many more passengers than a Boeing 767-300 can a Boeing 747-400 carry?

15. A jogger ran 18 miles in Week 1, 23 miles in Week 2, and 27 miles in Week 3. The jogger's goal is to run 100 miles a month. About how many miles will the jogger need to run during Week 4 to make the goal?

Spiral Review

16. Round $2.65 to the nearest dime.

17. Write the place value of the 5 in 345,678.

21

Warm-Ups

Standardized Testing Warm-Ups

1. You and three friends spent $.87, $.34, $.97, and $.71 at the store. About how much did you spend together?

    A $3.10   **B** $3.00   **C** $2.90   **D** $2.80

2. Which of the following amounts does *not* round to 600 to the nearest 100?

    A 567   **B** 619   **C** 550   **D** 652

Homework Review Warm-Ups

Round each number to the indicated place.

3. 539 (hundreds)   **4.** 608 (hundreds)   **5.** 970 (hundreds)

6. 7419 (thousands)   **7.** 3.8 (whole number)   **8.** 14.1 (whole number)

Project

1. In Lesson 1, you estimated the regular price for each item you would like to purchase for the school store. Select a reasonable selling price for each item. Then subtract the purchase cost from the selling price to estimate your profit for each item. Complete a table like the one below for the items you selected.

Item	Estimated selling price per item	Estimated cost per item	Estimated profit per item

2. You decide to package and sell "toolkits," made up of three or four single items. Choose items to put in your toolkit. Estimate the total cost, the total selling price, and the total profit for each toolkit.

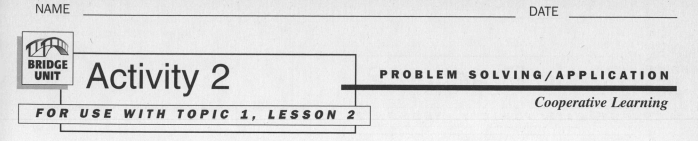

Activity 2

PROBLEM SOLVING/APPLICATION

Cooperative Learning

Estimating Costs

In this activity, you will explore using estimated sums in everyday life.

You will need:

- newspaper inserts or other advertisements for grocery stores
- 10 or 12 small slips of paper that are the same size
- plain paper
- a calculator (optional)

What to do:

Work with a partner.

1. From the ads, find the cost per pound of 10 or 12 different fruits or vegetables. Try to select items whose prices are not all real close together. On each slip of paper, write the price for one pound of a particular fruit or vegetable.

2. Fold and mix up the slips of paper. Then draw five of them without looking. Write down the prices on a sheet of paper.

3. Suppose you go into the store with a grocery list for one pound of each of the five items. Using estimation and mental math to total the prices, what is the smallest number of dollars you can have to be able to pay for the produce? What is your estimated price?

4. Return the slips, mix them up, and repeat the activity at least three more times.

5. Write answers to the following questions.

   a. How does the number of dollars you need relate to your estimated total price?

   b. What strategies did you use to find the total price?

   c. Did your partner ever find a different estimated total price? If so, how did this happen?

   d. Check each total price using a calculator or paper and pencil. Make sure that the number of dollars you estimated is enough to pay for the items.

Basic Practice 2

Estimate each sum or difference by rounding each number to the place held by the first digit.

1. 39 + 47   **2.** 52 + 85   **3.** 97 − 41

4. 82 + 54 + 61   **5.** 88 + 74 + 16   **6.** 167 − 89

7. 368 + 529   **8.** 887 − 295   **9.** 472 + 818

10. 506 − 283   **11.** 716 − 695   **12.** 246 + 468 + 802

13. 1278 + 4754   **14.** 3598 + 9216   **15.** 7802 − 1965

16. 4025 − 3416   **17.** 7502 − 6690   **18.** 1196 + 7812 + 6513

19. 2.3 + 8.9   **20.** 6.6 − 5.9   **21.** 8.2 − 7.3

22. 3.5 + 7.4 + 1.6 **23.** 2.1 + 8.8 + 3.6 **24.** 4.5 − 0.6

In each row, find three numbers whose sum is about 1000.

25. 243   124   567   819   314

26. 262   497   315   124   223

27. 315   483   287   408   389

Use the graph, which shows the weights of several animals in a zoo, to answer each question.

28. About how many more pounds does the polar bear weigh than the llama?

29. About how many more pounds does the llama weigh than the mountain lion?

30. Which animal weighs about 200 pounds more than the porpoise?

31. Would two wild boars the size given weigh about as much as the polar bear? Explain.

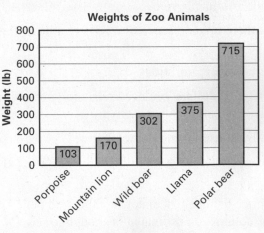

Weights of Zoo Animals

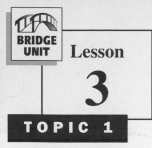

Number Patterns in Multiplication and Division

Goal
Convert metric units of length. Identify and use patterns involving multiplying and dividing by 10, 100, and 1000.

Steve jogged in a 10-kilometer race around the center of town. Each kilometer is made up of 1000 meters, so Steve jogged 10 × 1000 = 10,000 meters in the race.

Terms to Know	Example / Illustration
Metric System a measurement system that includes the length or distance measures of centimeters, meters, and kilometers	100 centimeters = 1 meter 1000 meters = 1 kilometer
Centimeter metric system unit of length or distance	There are a little more than 2.5 centimeters in an inch. in. 1 2 cm 1 2 3 4 5
Meter metric system unit of length or distance	A meter is a little longer than a yard. 1 m 1 yd
Kilometer metric system unit of length or distance	A distance of 10 kilometers is a little over 6 miles. Newburg 21 km (13 mi) Fairport 100 km (62 mi)

(continued)

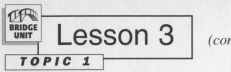

Lesson 3 *(continued)*

UNDERSTANDING THE MAIN IDEAS

By finding patterns, you can use mental math to multiply or divide by 10, 100, or 1000. Remember that the answer in a multiplication problem is the *product*, and the answer in a division problem is the *quotient*.

Example 1

Use mental math to find the products or quotients.

a. 4×10
4×100
4×1000

b. $4000 \div 10$
$4000 \div 100$
$4000 \div 1000$

▪ Solution ▪

Think of a whole number as having a decimal point to its right. You can think of 4 as 4.0 or of 4000 as 4000.0.

a. $4 \times 10 = 40$
$4 \times 100 = 400$
$4 \times 1000 = 4000$

b. $4000 \div 10 = 400$
$4000 \div 100 = 40$
$4000 \div 1000 = 4$

Notice the pattern: To multiply by 10, move the decimal one place to the right; to multiply by 100, move the decimal two places to the right; and so on.

Notice the pattern: To divide by 10, move the decimal one place to the left; to divide by 100, move the decimal two places to the left; and so on.

Find each product or quotient.

1. 8×100
2. 9×1000
3. 23×10

4. 41×100
5. 45×100
6. 57×1000

7. $60 \div 10$
8. $600 \div 10$
9. $8000 \div 10$

10. $7000 \div 100$
11. $3000 \div 1000$
12. $15,000 \div 1000$

A meter is larger than a centimeter. If you change a measure from meters to centimeters, the number will *increase*. To convert meters to centimeters, *multiply* the amount by 100.

A centimeter is smaller than a meter. If you change a measure from centimeters to meters, the number will *decrease*. To convert centimeters to meters, *divide* the amount by 100.

(continued)

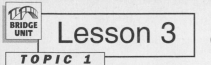

Lesson 3 | *(continued)*

Example 2

A reptile exhibit includes an Eastern diamond-backed rattlesnake that is 157 cm long and a mamba that is 1.5 m long. Which snake is longer?

■ Solution ■

To compare the lengths, they must use the same units of measure. You can either (1) write both lengths in centimeters or you can (2) write them both in meters.

Step 1: Write both lengths in centimeters.

The rattlesnake measures 157 cm.

To convert the length of the mamba from meters to centimeters, *multiply* by 100. You can do this mentally by moving the decimal point *two* places to the *right*.

1.5 m × 100 = 150 cm

Step 2: Write both lengths in meters.

The mamba measures 1.5 m.

To convert the length of the rattlesnake from centimeters to meters, *divide* by 100. You can do this mentally by moving the decimal point *two* places to the *left*.

157 cm ÷ 100 = 1.57 m

Because 157 > 150 and 1.57 > 1.5, the rattlesnake is longer than the mamba.

Use mental math to complete.

13. 5 m = __?__ cm

14. 2.7 m = __?__ cm

15. 32.17 m = __?__ cm

16. 700 cm = __?__ m

17. 143 cm = __?__ m

18. 15 cm = __?__ m

A kilometer is larger than a meter. If you change a measure from kilometers to meters, the number will *increase*. To convert kilometers to meters, *multiply* the amount by 1000.

A meter is smaller than a kilometer. If you change a measure from meters to kilometers, the number will *decrease*. To convert meters to kilometers, *divide* the amount by 1000.

(continued)

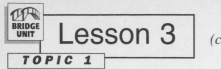

Lesson 3 *(continued)*

Example 3

The town of Cray is 3.1 km from Breyer and 3000 m from Abbott. Is Cray closer to Breyer or to Abbott?

▪ Solution ▪

To compare the lengths, they must use the same units of measure. You can either (1) write both lengths in meters or you can (2) write them both in kilometers.

Step 1: Write both lengths in meters.

Cray is 3000 m from Abbott.

To convert the distance from Cray to Breyer from kilometers to meters, *multiply* by 1000. You can do this mentally by moving the decimal point *three* places to the *right*.

$$3.1 \text{ km} \times 1000 = 3100 \text{ m}$$

Step 2: Write both lengths in kilometers.

Cray is 3.1 km from Breyer.

To convert the distance from Cray to Abbott from meters to kilometers, *divide* by 1000. You can do this mentally by moving the decimal point *three* places to the *left*.

$$3000 \text{ m} \div 1000 = 3 \text{ km}$$

Because 3000 < 3100 and 3 < 3.1, Cray is closer to Abbott than to Breyer.

Use mental math to complete.

19. 12 km = __?__ m

20. 1.6 km = __?__ m

21. 8.3 km = __?__ m

22. 7000 m = __?__ km

23. 13,000 m = __?__ km

24. 2400 m = __?__ m

⋯⋯⋯⋯⋯⋯

Spiral Review

25. Round 534.7 to the nearest whole number.

26. Of the 6724 pennies Roland collected, 2884 were dated 1990 or later. To the nearest thousand, how many were from 1989 or earlier?

27. Use estimation to decide if the $5 in your pocket is enough to pay for the items in your basket costing $1.87, $.74, $1.95, and $.81. Check your answer.

Bridge Unit, PASSPORT TO MATHEMATICS BOOK 1

Warm-Ups

FOR USE WITH TOPIC 1, LESSON 3

Standardized Testing Warm-Ups

1. What number should come next in the pattern below?

   ten, one hundred, one thousand, ten thousand, one hundred thousand…

   A one million   **B** ten million

   C one hundred million   **D** not here

2. Which pair contains measurements that describe the same length?

   A 2.5 km and 250 m   **B** 180 cm and 1.8 m

   C 20 cm and 2000 m   **D** 3600 m and 36 km

Homework Review Warm-Ups

Estimate each sum or difference by first rounding each number to the place held by the first digit and then using mental math.

3. 465 + 719   4. 3129 + 6913

5. 847 − 229

6. A fast-food chain mails 4782 coupons for customers to sample a new salad. Of these coupons, 3098 are used. Use estimation to the nearest hundred to find about how many coupons are not used.

Project

FOR USE WITH TOPIC 1, LESSON 3

Suppose an empty classroom the same size as yours is chosen to be the school store. For you to design the store, including shelf space, displays, a purchasing area, and decorations, you will need to know the size of the room.

Measure the length and width of your classroom in metric units. Make your measurements to the nearest hundredth of a meter. Record these lengths and then convert them to centimeters. How did you make the conversions?

Is it easier for you to understand or picture the room measurements in meters or centimeters? Tell why you think so.

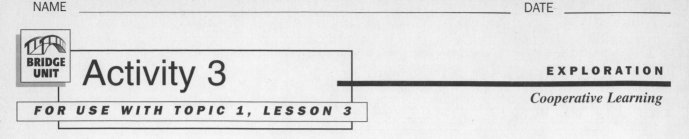

Activity 3

FOR USE WITH TOPIC 1, LESSON 3

An Estimating Game: Target 100,000

In this activity, you will try to get closer to a sum of 100,000 than your partners *without going over.*

You will need:

- index cards
- plain paper
- a calculator (optional)

What to do:

Work in groups of 3–5. Work with your partners to make game cards from the index cards. There are 18 game cards in all, each with one number written on it. Make two cards of each number: 11.1, 22.2, 33.3, 44.4, 55.5, 66.6, 77.7, 88.8, and 99.9. Then begin the game by shuffling the cards and placing them face down between players.

1. Take turns choosing one card. Multiply the number you see by 10, 100, or 1000. Write the product on your own piece of paper.

2. After all players have taken four turns, find each player's estimated sum by rounding each product to the nearest 1000. Record a preliminary ranking of the sums by ordering them from the closest to 100,000 to the farthest from 100,000.

3. To find the final ranking, find the actual sums and subtract them from 100,000. (Use a calculator if you wish.) The player whose actual difference from 100,000 is *smallest* wins the round. Any player whose actual score is over 100,000 is disqualified from the round.

4. After playing two rounds of Target 100,000, answer the following questions.

    a. What mental math did you use to help you decide whether to multiply by 10, 100, or 1000?

    b. Describe how you estimated the sum of your numbers.

    c. Did you use a different strategy in the second round than in the first round? If you did, explain how it was different.

Basic Practice 3

FOR USE WITH TOPIC 1, LESSON 3

Find each product or quotient.

1. 5×10 **2.** 19×10 **3.** 37×100

4. 451×100 **5.** 35×1000 **6.** 627×1000

7. $900 \div 10$ **8.** $7000 \div 10$ **9.** $8000 \div 100$

10. $19,000 \div 10$ **11.** $35,000 \div 1000$ **12.** $315,000 \div 100$

Use mental math to complete Exercises 13–28.

13. 6 m = __?__ cm **14.** 7.9 m = __?__ cm

15. 32 m = __?__ cm **16.** 7300 cm = __?__ m

17. 1600 cm = __?__ m **18.** 1459 m = __?__ cm

19. 234,000 cm = __?__ m **20.** 1,000,000 cm = __?__ m

21. 2.5 km = __?__ m **22.** 5.96 km = __?__ m

23. 6200 m = __?__ km **24.** 10,000 m = __?__ km

25. 2000 m = __?__ km **26.** 16 km = __?__ m

27. 1,000,000 m = __?__ km **28.** 1 km = __?__ m

29. Write 24,000 as a product of 10 and a number, of 100 and a number, of 1000 and a number, and of 10,000 and a number.

30. Write 1200 in three different ways as the product of 100 times two different whole numbers.

31. Tanya is planning a hike. Her trail guide lists the Meadowbrook Trail as 1.7 km long and the Larkspur Trail as 800 m long. If she hikes both trails, what is the total distance she hikes? Give your answer both in kilometers and in meters.

32. Antonio says that it is about 75,000,000 cm from his house to the nearest ocean beach. If the distance is 750 km, is he right? Explain.

BRIDGE UNIT

Lesson

4

TOPIC 1

Goal
Use rounding to estimate products. Use estimation to make decisions and solve problems.

Estimating Products

In Lesson 1, you learned how to round numbers. In Lesson 3, you found a pattern to find products of 10, 100, and 1000. You can use rounding and patterns to estimate the product of any two numbers.

UNDERSTANDING THE MAIN IDEAS

Estimating a product can help you make a decision when you do not need an exact answer. It can also help you determine if an answer is reasonable.

Example 1

A common hippopotamus eats only plants. Each night, a hippopotamus can eat about 38 kilograms of grass. Estimate the number of kilograms of grass a hippopotamus can eat in a week. (One kilogram is a metric system measure that equals about 2.2 pounds—there are 1000 grams in a kilogram.)

■ Solution ■

Step 1: Use rounding to estimate the amount of grass a hippopotamus eats in a day.

Round 38 up to 40.

Step 2: Multiply the estimated amount by the number of days in a week.

$7 \times 40 = 280$

A hippopotamus can eat about 280 kilograms of grass in one week.

Round the first number in each pair to the place held by the first digit to estimate each product.

1. 53×8 **2.** 33×9 **3.** 721×3 **4.** 869×7 **5.** 231×4

(continued)

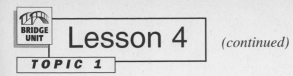

Remember that *factors* are the numbers you multiply together to find a product. You can use rounding and patterns to estimate products when both factors are greater than 10. First notice the pattern shown below.

$4 \times 6 = 24$

$4 \times 6\underline{0} = 24\underline{0}$

$4\underline{00} \times 6 = 24\underline{00}$

$4\underline{0} \times 6\underline{00} = 24,\underline{000}$

$4\underline{00} \times 6\underline{00} = 24\underline{0,000}$

$4\underline{0} \times 6 = 24\underline{0}$

$4\underline{0} \times 6\underline{0} = 24\underline{00}$

$4 \times 6\underline{00} = 24\underline{00}$

$4\underline{00} \times 6\underline{0} = 24,\underline{000}$

To find how many zeros follow the product of 4 and 6, *add the number of zeros in each factor.*

Example 2

Jarrod orders 32 cases of trading cards for his store. Each case contains 48 packs of cards. About how many packs of trading cards will Jarrod have to sell?

■ Solution ■

Step 1: Use rounding to estimate each factor.

Round 32 down to 30.

Round 48 up to 50.

Step 2: Multiply the rounded numbers to estimate the product.

$3\underline{0} \times 5\underline{0} = 15\underline{00}$

Because each factor has 1 zero, the result has $1 + 1 = 2$ zeros following the product of 3 and 5.

Jarrod will have about 1500 packs of trading cards to sell.

Estimate each product by rounding the numbers in each pair to the place held by the first digit.

6. 41×18 **7.** 78×12 **8.** 56×83 **9.** 215×37 **10.** 673×24

11. Evan thinks the product of 28 and 28 will be over 1000. Do you agree? Explain.

(continued)

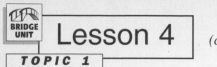

You can use rounding to estimate products involving measures of time.

Example 3

Jennifer spends about 1 hour and 55 minutes on the subway each day traveling to and from work. About how long does Jennifer spend traveling in a 5 day work week?

■ **Solution** ■

Step 1: Round the time to the nearest hour. Remember that there are 60 minutes in an hour.

Because 55 minutes is close to 1 hour, 1 hour and 55 minutes is close to 2 hours.

Step 2: Multiply the number of hours spent traveling per day by the number of days to estimate the total travel time.

$2 \times 5 = 10$

Jennifer spends about 10 hours a week traveling to and from work.

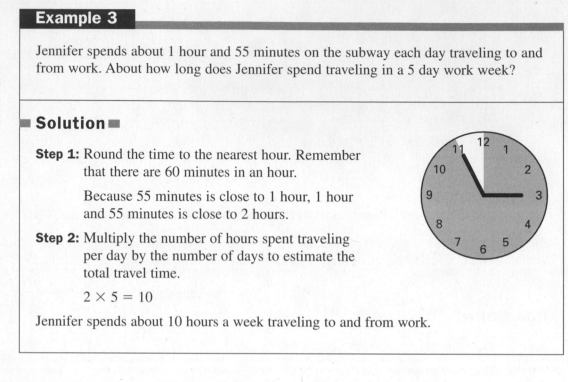

Round each amount of time to the nearest hour. Then multiply to estimate the total time.

12. Arturo works 8 hours and 45 minutes each day, 6 days a week. Estimate the number of hours Arturo works each week.

13. Martha's hourly wage doubles once she has worked more than 40 hours in one week. This week, she worked 9 hours and 10 minutes each day for 5 days. Did she work enough to make overtime pay? Explain.

14. You practice piano 50 minutes a day. Is it reasonable to say that you practice more than one whole day a month? Explain.

· · · · · · · · · · · · · · · · · · · ·
Spiral Review

15. Estimate the sum of 2.3, 3.4, and 4.8.

16. It is 2.5 km from Brendan's home to the school. If each pace Brendan walks is about 1 meter, how many paces will he take to walk from home to school?

Warm-Ups

FOR USE WITH TOPIC 1, LESSON 4

Standardized Testing Warm-Ups

1. Which number does *not* divide evenly by 100?

A 17,000   **B** 1400   **C** 9500   **D** 10,010

2. Without multiplying, tell which one of the following products could *not* be true.

A $21 \times 20 = 410$ **B** $11 \times 610 = 6710$

C $110 \times 300 = 3300$ **D** $440 \times 220 = 96,800$

Homework Review Warm-Ups

A school supply company has created four new products for this school year. The table shows how many of each you have ordered for the school store.

3. Use estimation to decide for which new products the number of items ordered is greater than 5000.

4. Find the number of items ordered for each new product.

New Products

	Number per carton	Number of cartons
A	198	10
B	45	100
C	60	100
D	72	1000

Project

FOR USE WITH TOPIC 1, LESSON 4

In Lesson 2, you estimated the profit earned for selling each item at the school store. To estimate your *total* profit for sales of an item, multiply the profit per item by the number of items ordered. Use the quantities in the table below to estimate the total profit for sales of each item you listed in the Lesson 2 Project table.

Item	Estimated profit per item	Number of items ordered	Estimated total profit
		28	
		36	
		42	
		50	
		60	

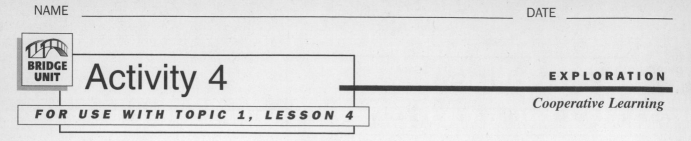

Activity 4

EXPLORATION

Cooperative Learning

An Estimating Game: Over and Under

In this activity, you will estimate products to try to go over and under a target number.

You will need:

- four number cubes: two labeled 0–5 and two labeled 4–9 (You can use the template below to make cubes, using construction paper and tape. Or, you can renumber many existing cubes using a pencil, and then erase the marks when you are done.)
- a calculator

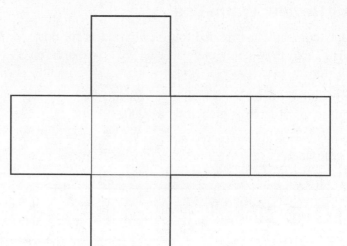

What to do:

Work with a partner.
  A. Choose one player to go first.
  - The first player's goal is to construct estimated products *under* 2500.
  - The second player's goal is to construct estimated products *over* 2500.

  B. Players take turns. Roll all four number cubes.

  C. Use the roll to form two 2-digit numbers (a number such as 06 is not allowed). Each player chooses the numbers so that the estimated product fits his or her "under" or "over" goal. Choose your numbers as quickly as you can.

  D. Use a calculator to find the actual product. If the actual product meets the player's goal, he or she scores 1 point. Switch goals, roll again, and repeat the game. The first player to score 5 points wins the round. (If you reach 5 points at the same time, keep rolling until one of you is ahead).

After two rounds of play, record your answers to the following questions.

  1. Describe your strategy for choosing the 2-digit numbers to meet your goal.

  2. Tell how your strategies for finding products under and over 2500 are different.

  3. Describe how you estimated the product of each number pair.

Basic Practice 4

FOR USE WITH TOPIC 1, LESSON 4

In Exercises 1–20, use rounding to estimate each product.

1. 27×9	**2.** 46×5	**3.** 79×9	**4.** 32×3
5. 17×7	**6.** 314×8	**7.** 819×4	**8.** 908×6
9. 3318×7	**10.** 1987×3	**11.** 52×28	**12.** 72×17
13. 81×68	**14.** 49×92	**15.** 97×11	**16.** 112×58
17. 26×12	**18.** 30×31	**19.** 51×42	**20.** 19×99
21. 215×28	**22.** 632×47	**23.** 189×33	**24.** 607×69

25. There are 5280 feet in a mile and 12 inches in a foot. Explain how you can tell quickly if there are more or fewer than 50,000 inches in a mile.

26. Which of the following times (there may be more than one) would you round to "about 4 hours?" Explain your answer.

    a. 3 hours, 19 minutes   **b.** 3 hours, 35 minutes

    c. 4 hours, 18 minutes   **d.** 4 hours, 41 minutes

27. If there are 48 eggs in a box and 42 boxes of eggs, about how many eggs are there all together?

Use the table at the right to answer Exercises 28–29.

28. Estimate the number of hours each employee will work this week.

29. Employees who work more than 40 hours this week can enter a drawing for a new bike. Which employee(s) can enter the drawing?

SoundYard Work Schedulle

Name	Hours	Days
Stephanie	5 h, 10 min	5
Jon	8 h, 15 min	4
Isaiah	7 h, 45 min	6
Suzanne	9 h, 5 min	4

Goal

Use rounding and
compatible numbers
to estimate quotients.
Use estimated
quotients to make
decisions and solve
problems.

Estimating Quotients

In the late 1700s and early 1800s, people often traveled by stagecoach.
A stagecoach could travel about 10 miles per hour. You can use rounding and
compatible numbers to estimate the time it would take to travel the 243 miles
from Dallas to Houston by stagecoach at that speed.

Terms to Know	*Example / Illustration*
Compatible numbers numbers that are easy to work with mentally; You use them in place of actual numbers to make an estimate, especially when dividing.	To estimate $243 \div 10$, round 243 to 240. $240 \div 10 = 24$ The numbers 240 and 10 are compatible because 10 divides 240 with a remainder of 0.
Dividend the number that is divided in a division problem	In $240 \div 10$, the dividend is 240.
Divisor the number by which you divide the dividend in a division problem	In $240 \div 10$, the divisor is 10.
Quotient the answer in a division problem	For $240 \div 10$, the quotient is 24.

UNDERSTANDING THE MAIN IDEAS

When you do not need an exact answer, you can estimate a quotient. This can
help you make decisions or decide if an answer is reasonable.

(continued)

Lesson 5 *(continued)*

Example 1

A wood strip is 228 cm long. Ms. Landy needs to cut the wood strip into 6 equal pieces. Estimate the length of each piece of wood.

■ Solution ■

Step 1: Choose numbers close to 228 that are multiples of 6. Think about basic multiplication facts to help you.

$$6 \times 2 = 12 \qquad 6 \times 3 = 18 \qquad 6 \times 4 = 24$$

Now you can see that the products below must also be true.

$$6 \times 20 = 120 \qquad 6 \times 30 = 180 \qquad 6 \times 40 = 240$$

228

<--+-------+-------+-------+-------+-------+--●----+-->
  120   140   160   180   200   220   240

Because 228 is closer to 240 than to 120 or 180, round 228 up to 240.

Step 2: Now divide to estimate the quotient.

$$240 \div 6 = 40$$

Each piece of wood will be about 40 cm long.

Use compatible numbers to estimate each quotient.

1. $147 \div 3$
2. $556 \div 7$
3. $636 \div 8$
4. $475 \div 6$
5. $1343 \div 4$
6. $7141 \div 9$
7. $6279 \div 9$
8. $5598 \div 6$

When the divisor is greater than 10, you may need to round both the dividend and the divisor to make them compatible.

(continued)

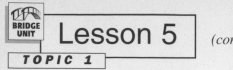

Example 2

Amy's highway speed averages about 58 miles per hour. About how many hours should she plan for a 317 mile trip?

■ Solution ■

Step 1: To estimate the quotient $317 \div 58$, round the divisor first.

Round 58 up to 60.

Step 2: Use basic facts and multiples to find a number close to 317 that is also compatible with 60.

$$4 \times 60 = 240 \qquad 5 \times 60 = 300 \qquad 6 \times 60 = 360$$

317

| | | | | | | |
|240|260|280|300|320|340|360|

Because 317 is closer to 300 than to 240 or 360, round 317 down to 300.

Step 3: Now divide to estimate the quotient.

$$300 \div 60 = 5$$

Amy should plan on about 5 hours for her trip.

In Exercises 9–16, write the compatible numbers you would use to estimate the quotient. Then write the estimate.

9. $456 \div 21$ **10.** $342 \div 56$ **11.** $895 \div 32$ **12.** $1003 \div 54$

13. $1268 \div 42$ **14.** $1421 \div 29$ **15.** $3567 \div 68$ **16.** $2093 \div 72$

17. An airplane's flight length is 163 minutes. The crew can choose to show a film that runs about 2 hours or one that runs about 3 hours. Which film should they select for the flight? Explain.

·········· Spiral Review

18. Estimate the product of 478 and 32.

19. Pat thinks that five markers costing $3.75 each will cost about $15. Ed thinks they will cost about $20. What do *you* think? Explain.

Warm-Ups

FOR USE WITH TOPIC 1, LESSON 5

Standardized Testing Warm-Ups

1. Which number has a 4 in the hundred-thousands' place?

  A 123,456   **B** 345,678   **C** 456,789   **D** 789,234

2. Which of the following numbers is *not* a multiple of 70?

  A 350   **B** 3500   **C** 420   **D** 670

Homework Review Warm-Ups

Estimate each product by rounding each number to the place held by its first digit.

3. 71×9 **4.** 84×78 **5.** 325×57

Project

FOR USE WITH TOPIC 1, LESSON 5

1. The school store will be open for business during lunch periods.

  a. How many lunch periods are there in your school?

  b. Estimate the number of minutes in each lunch period.

  c. Use the information in parts (a) and (b) to estimate the total number of minutes that the store will be open each month. Assume that there are four weeks in a month.

2. The students in your class will be the clerks for the store. The clerks will work one at a time, and each clerk will work the same amount of time. Estimate the number of minutes each student will need to work per month.

3. Describe how you estimated the number of minutes each student would work per month. Did you use rounding? compatible numbers?

4. With a calculator or paper and pencil, find the actual amount of time each student would need to work each month. How does your estimate compare?

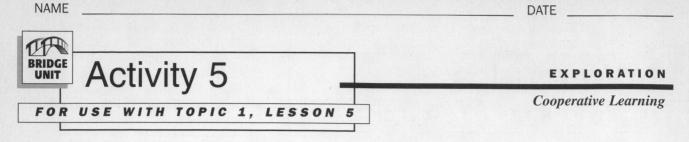

Quick Quotient Buster

In this activity, you'll see how you can use a special basic facts chart to estimate quotients.

You will need:

- a blank number grid
- plain paper

What to do:

Work with a partner.

1. Use the blank number grid to make a special multiplication table. Write a multiplication sign in the top left corner box. Write the numbers 1–10 across the top of the grid. Write the multiples of 10 from 10 through 100 down the left side. Complete the table by filling in the products of these numbers.

×	1	2	3	4	. . .
10	10	20	30		
20	20				
30	30				
40			120		
50				200	
60		120			
. . .					

2. Use the completed table to estimate each quotient.

   a. 195 ÷ 60   **b.** 256 ÷ 40   **c.** 478 ÷ 70   **d.** 536 ÷ 80

3. How did you use the table to estimate the quotients?

4. Suppose you want to estimate the quotient 3743 ÷ 72. Describe a table that would help you solve the problem. Then tell how you would use the table.

Basic Practice 5

Use rounding and compatible numbers to estimate each quotient.

1. $832 \div 4$	**2.** $152 \div 7$	**3.** $267 \div 5$	**4.** $711 \div 8$
5. $312 \div 5$	**6.** $171 \div 2$	**7.** $466 \div 8$	**8.** $563 \div 6$
9. $889 \div 9$	**10.** $592 \div 6$	**11.** $782 \div 2$	**12.** $904 \div 3$
13. $8215 \div 4$	**14.** $6321 \div 9$	**15.** $7980 \div 8$	**16.** $5000 \div 7$
17. $3247 \div 8$	**18.** $2697 \div 3$	**19.** $1495 \div 5$	**20.** $1999 \div 4$
21. $178 \div 18$	**22.** $138 \div 23$	**23.** $627 \div 28$	**24.** $163 \div 17$
25. $6216 \div 68$	**26.** $7210 \div 32$	**27.** $2470 \div 53$	**28.** $4900 \div 13$
29. $5780 \div 74$	**30.** $2650 \div 31$	**31.** $1738 \div 64$	**32.** $3469 \div 43$
33. $3456 \div 57$	**34.** $8200 \div 17$	**35.** $5575 \div 73$	**36.** $7709 \div 78$
37. $4970 \div 68$	**38.** $2350 \div 75$	**39.** $6321 \div 34$	**40.** $4128 \div 82$

Use the table for Exercises 41 and 42.
To find time, divide distance by speed.

41. Suppose a jet flies at a speed of about 600 miles per hour. Which cities would be less than 3 hours flying time from Los Angeles, California?

42. Lucio wants to drive from Los Angeles, California to Dallas, Texas. At an average speed of 58 miles per hour, about how long will it take him to get there?

43. A flight attendant says today's flight time from Los Angeles to Atlanta will be 228 minutes. What is the estimated flight time in hours?

Distances from Los Angeles, California to Selected Cities

City, State	Distance (mi)
Atlanta, Georgia	2182
Boston, Massachusetts	2979
Chicago, Illinois	2084
Dallas, Texas	1387
Houston, Texas	1538
Milwaukee, Wisconsin	2087
Portland, Oregon	959
San Francisco, California	379

Basic Assessment

FOR USE WITH TOPIC 1

Round each number to the place held by the first digit.

1. 53	**2.** 87	**3.** 25	**4.** 76	**5.** 84
6. 229	**7.** 681	**8.** 190	**9.** 919	**10.** 650
11. 2345	**12.** 1798	**13.** 3612	**14.** 4246	**15.** 9501
16. 3.7	**17.** 9.2	**18.** 8.6	**19.** 4.5	**20.** 1.1

Estimate each sum or difference by rounding each number to the place held by the first digit.

21. $381 + 625$ **22.** $3091 + 2125$ **23.** $4.3 + 8.9$

24. $82 - 48$ **25.** $6389 - 5906$ **26.** $9.0 - 2.1$

Use mental math to find each product or quotient.

27. 35×100 **28.** 29×1000 **29.** $4000 \div 10$

30. 275×100 **31.** $3210 \div 10$ **32.** 5.7×1000

Estimate each product by rounding to the tens' digit. Use compatible numbers to estimate each quotient.

33. 43×91 **34.** 87×69 **35.** 31×63

36. $550 \div 72$ **37.** $4680 \div 47$ **38.** $2470 \div 53$

Solve.

39. Arnie's stride is about 1 meter long. If he runs a 10 kilometer race, about how many paces will he travel?

40. An ice cream recipe calls for 2350 milliliters of milk. How many liter containers of milk must you buy in order to make the recipe? (Milliliters and liters are metric system measures of volume; 1000 milliliters = 1 liter. A liter is a little more than a quart.)

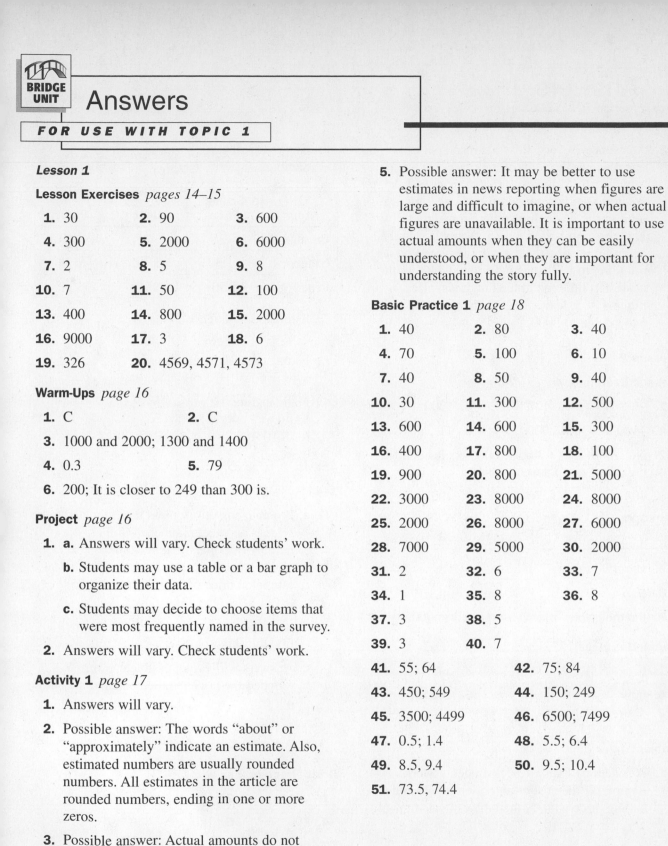

Lesson 1

Lesson Exercises *pages 14–15*

1. 30	**2.** 90	**3.** 600
4. 300	**5.** 2000	**6.** 6000
7. 2	**8.** 5	**9.** 8
10. 7	**11.** 50	**12.** 100
13. 400	**14.** 800	**15.** 2000
16. 9000	**17.** 3	**18.** 6
19. 326	**20.** 4569, 4571, 4573	

Warm-Ups *page 16*

1. C **2.** C

3. 1000 and 2000; 1300 and 1400

4. 0.3 **5.** 79

6. 200; It is closer to 249 than 300 is.

Project *page 16*

1. a. Answers will vary. Check students' work.

b. Students may use a table or a bar graph to organize their data.

c. Students may decide to choose items that were most frequently named in the survey.

2. Answers will vary. Check students' work.

Activity 1 *page 17*

1. Answers will vary.

2. Possible answer: The words "about" or "approximately" indicate an estimate. Also, estimated numbers are usually rounded numbers. All estimates in the article are rounded numbers, ending in one or more zeros.

3. Possible answer: Actual amounts do not usually end in zero.

4. Answers will vary.

5. Possible answer: It may be better to use estimates in news reporting when figures are large and difficult to imagine, or when actual figures are unavailable. It is important to use actual amounts when they can be easily understood, or when they are important for understanding the story fully.

Basic Practice 1 *page 18*

1. 40	**2.** 80	**3.** 40
4. 70	**5.** 100	**6.** 10
7. 40	**8.** 50	**9.** 40
10. 30	**11.** 300	**12.** 500
13. 600	**14.** 600	**15.** 300
16. 400	**17.** 800	**18.** 100
19. 900	**20.** 800	**21.** 5000
22. 3000	**23.** 8000	**24.** 8000
25. 2000	**26.** 8000	**27.** 6000
28. 7000	**29.** 5000	**30.** 2000
31. 2	**32.** 6	**33.** 7
34. 1	**35.** 8	**36.** 8
37. 3	**38.** 5	
39. 3	**40.** 7	

41. 55; 64	**42.** 75; 84
43. 450; 549	**44.** 150; 249
45. 3500; 4499	**46.** 6500; 7499
47. 0.5; 1.4	**48.** 5.5; 6.4
49. 8.5, 9.4	**50.** 9.5; 10.4
51. 73.5, 74.4	

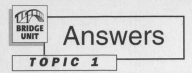
52. Answers will vary. Possible response: Look one place to the right of the place you are rounding to. If the digit in that place is 5 or greater, then you round up by increasing the digit in the place that you are rounding to by 1. If the digit is less than 5, then you round down by leaving the digit in the place you are rounding to unchanged. To round 3468 to the nearest 1000, look at the hundreds' place. Since $4 < 5$, leave the 3 unchanged and round 3468 down to 3000.

Lesson 2

Lesson Exercises *pages 20–21*

1. 90   **2.** 130   **3.** 1200

4. 1100   **5.** 9000   **6.** 11

7. No; about $200 + 400 = 600$ people attended on Friday and Saturday.

8. 40   **9.** 200   **10.** 400

11. 3000   **12.** 5   **13.** 0

14. about 200

15. about 30 miles

16. $2.70

17. the thousands' place (It represents 5000.)

Warm-Ups *page 22*

1. C   **2.** D   **3.** 500

4. 600   **5.** 1000   **6.** 7000

7. 4   **8.** 14

Project *page 22*

1. Answers will vary. Check students' work. Sample: If the cost of an eraser is $.25 and the selling price is $.35, then the estimated profit per eraser is $.10.

2. Answers will vary. Sample:

Item	Selling Price	Cost	Profit
Eraser	$.35	$.25	$.10
Notebook	$1.50	$.79	$.71
Pencil	$.50	$.29	$.21

Estimated toolkit cost:
$.30 + $.80 + $.30 = $1.40

Estimated toolkit selling price:
$.40 + $1.50 + $.50 = $2.40

Estimated toolkit profit:
$2.40 − $1.40 = $1.00
(or $.10 + $.70 + $.20 = $1.00)

Activity 2 *page 23*

1–4. Answers will vary.

5. a. Possible answer: Unless the estimated total price is an exact number of dollars, the number of dollars you need is the next higher whole number. For example, for an estimated price of $4.10, you need $5.

b. Possible answer: I used rounding to the nearest dime, and used paper and pencil to find the sum.

c. Answers will vary. Possible answer: Yes; I rounded the prices to the nearest dime, and my partner rounded them to the nearest dollar.

d. Answers will vary.

Basic Practice 2 *page 24*

1. 90   **2.** 140   **3.** 60

4. 190   **5.** 180   **6.** 80

7. 900   **8.** 600   **9.** 1300

10. 200   **11.** 0   **12.** 1500

13. 6000   **14.** 13,000   **15.** 6000

16. 1000   **17.** 1000   **18.** 16,000

19. 11   **20.** 1   **21.** 1

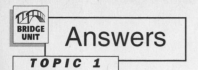
22. 13   **23.** 15   **24.** 4

25. One possible solution: 124, 567, and 314

26. One possible solution: 262, 497, and 223

27. One possible solution: 315, 287, and 408

28. about 300 pounds

29. about 200 pounds

30. the wild boar

31. No; the 2 wild boars would weigh only about 600 pounds, but the polar bear weighs about 700 pounds.

Lesson 3

Lesson Exercises *pages 26–28*

1. 800   **2.** 9000   **3.** 230

4. 4100   **5.** 4500   **6.** 57,000

7. 6   **8.** 60   **9.** 800

10. 70   **11.** 3   **12.** 15

13. 500   **14.** 270   **15.** 3217

16. 7   **17.** 1.43   **18.** 0.15

19. 12,000   **20.** 1600   **21.** 8300

22. 7   **23.** 13   **24.** 2.4

25. 535   **26.** about 4000

27. No; the exact cost is $5.37.

Warm-Ups *page 29*

1. A   **2.** B   **3.** 1200

4. 10,000   **5.** 600   **6.** about 1700

Project *page 29*

Check students' work. Students are likely to say that it is easier to work with meters than with centimeters because the number of meters in the length and width will be less than 20, while the number of centimeters will be over 500.

Activity 3 *page 30*

1–3. Answers will vary.

4. a. Answers will vary. Students may say they estimated their subtotal, subtracted from 100,000, and used the difference to decide by which number to multiply.

    b. Answers will vary. Some students will say they rounded each number to its greatest place value, then added mentally. Others may say they used front-end estimation.

    c. Answers will vary. Students may say they tried to come as close as possible to 100,000 with the first turn, then used subsequent turns to "fill in."

Basic Practice 3 *page 31*

1. 50   **2.** 190   **3.** 3700

4. 45,100   **5.** 35,000   **6.** 627,000

7. 90   **8.** 700   **9.** 80

10. 1900   **11.** 35   **12.** 3150

13. 600   **14.** 790   **15.** 3200

16. 73   **17.** 16   **18.** 145,900

19. 2340   **20.** 10,000   **21.** 2500

22. 5960   **23.** 6.2   **24.** 10

25. 2   **26.** 16,000   **27.** 1000

28. 1000

29. 10×2400; 100×240; 1000×24; $10,000 \times 2.4$

30. $100 \times 1 \times 12$; $100 \times 2 \times 6$; $100 \times 3 \times 4$

31. 2.5 km; 2500 m

32. Yes; 75,000,000 cm is 750,000 m, and 750,000 m is 750 km.

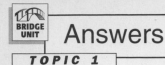
Lesson 4

Lesson Exercises *pages 32–34*

1. 400
2. 270
3. 2100
4. 6300
5. 800
6. 800
7. 800
8. 4800
9. 8000
10. 14,000

11. No; the product will be less than 900. Both numbers are rounded up to 30, and 30 × 30 = 900.

12. about 54 hours

13. Yes; Martha worked about 45 hours this week.

14. Yes; 50 minutes a day × 30 days is about 30 hours. There are only 24 hours in a day.

15. about 10
16. about 2500 paces

Warm-Ups *page 35*

1. D
2. C
3. C and D
4. A: 1980; B: 4500; C: 6000; D: 72,000

Project *page 35*

Answers will vary. Check students' work. Sample: If the estimated profit per pencil is $.21 and 28 pencils are ordered, then the estimated total profit is about $.20 × 30 = $6.

Activity 4 *page 36*

1. Possible answer: I looked for two number cubes whose product would be less than (greater than) 25; I used those for the tens places of the numbers.

2. Possible answer: To find products over 2500, I used the two largest numbers showing for each number's tens' place. For the under 2500 goal, I used the two smallest numbers showing for each number's tens' place, except when one or more cubes showed zero.

3. Possible answer: I rounded each 2-digit number to the nearest ten. Then, I multiplied and inserted zeros.

Basic Practice 4 *page 37*

1. 270
2. 250
3. 720
4. 90
5. 140
6. 2400
7. 3200
8. 5400
9. 21,000
10. 6000
11. 1500
12. 1400
13. 5600
14. 4500
15. 1000
16. 6000
17. 300
18. 900
19. 2000
20. 2000
21. 6000
22. 30,000
23. 6000
24. 42,000

25. Possible answer: Because 5280 is more than 5000, and 12 is more than 10, the number of inches must be more than 5000 × 10 = 50,000.

26. (b) and (c); Since 1 hour = 60 minutes, there must be 30 minutes to round up to the next hour. For (a), 3 hours, 19 minutes is rounded to 3 hours. For (d), 4 hours, 41 minutes is rounded to 5 hours.

27. about 2000

28. Stephanie: about 25 hours; Jon: about 32 hours; Isaiah: about 48 hours; Suzanne: about 36 hours

29. Isaiah

Lesson 5

Lesson Exercises *pages 39–40*

1. 50
2. 80
3. 80
4. 80
5. 300
6. 800
7. 700
8. 900
9. 400 ÷ 20; 20
10. 360 ÷ 60; 6
11. 900 ÷ 30; 30
12. 1000 ÷ 50; 20
13. 1200 ÷ 40; 30
14. 1500 ÷ 30; 50
15. 3500 ÷ 70; 50
16. 2100 ÷ 70; 30

17. The crew should choose the 2-hour film. Because 163 minutes is less than 3 hours, they will not have enough time to show the entire 3-hour film.

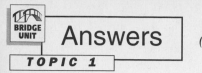

Answers *(continued)*

18. about 15,000

19. Possible answer: I think Ed is right. $3.75 is rounded to $4.00, and $4 \times 5 = 20$.

Warm-Ups *page 41*

1. C   **2.** D   **3.** about 630

4. about 6400   **5.** about 18,000

Project *page 41*

1–2. Check students' work. Sample: If there are two lunch periods per day and each is 45 minutes long, that is 90 minutes per day. This is $90 \times 5 \times 4$ (or 90×20) minutes per month, or about 1800 minutes per month. If there are 28 students in the class, each would work about $1800 \div 30 = 60$ minutes per month.

3–4. Answers will vary.

Activity 5 *page 42*

1. Check students' charts.

2. a. about 3

     b. about 6

     c. about 7

     d. about 7

3. Possible answer: Moving across the row showing the divisor, look for two numbers that the actual dividend lies between. Choose the number that is closer to the actual dividend. Move up to the top of the column to find the estimated quotient.

4. Possible answer: A table might be made showing the multiples from 10–100 across the top and down the side. Round 72 to 70. Look across the "70" row. Find the two numbers that 3743 lies between. Move up to the top of the column to find the estimated quotient (50).

Basic Practice 5 *page 43*

1. 200   **2.** 20   **3.** 50

4. 90   **5.** 60   **6.** 90

7. 60   **8.** 90   **9.** 100

10. 100   **11.** 400   **12.** 300

13. 2000   **14.** 700   **15.** 1000

16. 700   **17.** 400   **18.** 900

19. 500   **20.** 500   **21.** 9 or 10

22. 7   **23.** 20   **24.** 8

25. 90   **26.** 200   **27.** 50

28. 500   **29.** 80   **30.** 90

31. 30   **32.** 90   **33.** 60

34. 400   **35.** 80   **36.** 100

37. 70   **38.** 30   **39.** 200

40. 50

41. Dallas, Texas; Houston, Texas; Portland, Oregon; San Francisco, California

42. about 20 hours

43. about 4 hours

Topic 1 Basic Assessment *page 44*

1. 50   **2.** 90   **3.** 30

4. 80   **5.** 80   **6.** 200

7. 700   **8.** 200   **9.** 900

10. 700   **11.** 2000   **12.** 2000

13. 4000   **14.** 4000   **15.** 10,000

16. 4   **17.** 9   **18.** 9

19. 5   **20.** 1   **21.** 1000

22. 5000   **23.** 13   **24.** 30

25. 0   **26.** 7   **27.** 3500

28. 29,000   **29.** 400   **30.** 27,500

31. 321   **32.** 5700   **33.** 3600

34. 6300   **35.** 1800   **36.** 8

37. 90 or 100   **38.** 50

39. 10,000 paces   **40.** 3 containers

Goal

Apply rules for divisibility of numbers by 2, 3, 5, 9, and 10.

Divisibility

All of the numbers 2, 3, 5, 9, and 10 will divide evenly into 1260. To know this, you don't actually have to do the division. By looking for patterns, you can find rules that give you shortcuts for deciding if one number divides evenly into another.

Terms to Know	Example / Illustration
Divisible when a number can be divided into equal parts by another number with a remainder of 0	$$\begin{array}{r} 137 \\ 2\overline{)274} \\ 2 \\ \hline 7 \\ 6 \\ \hline 14 \\ 14 \\ \hline 0 \end{array}$$ ← remainder Since the remainder is 0, 274 is divisible by 2.
Divisibility rules rules that tell you whether or not one number can be evenly divided by another, smaller number	Numbers divisible by 2 have a pattern that you can see in the ones' place. The pattern can be used to make a rule about divisibility by 2.

UNDERSTANDING THE MAIN IDEAS

You probably know the divisibility rule for 2 already. You are using this rule whenever you say that a number is *even* or *odd*.

(continued)

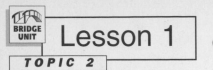

Lesson 1 *(continued)*

Example 1

Danny wants to know if 7138 is divisible by 2. Sharisse says she can tell that it is without dividing. How does Sharisse know that 7138 is divisible by 2?

■ Solution ■

A number is divisible by 2 if the digit in the ones' place is even. The even digits are 0, 2, 4, 6, and 8.

Notice that 7138 has
an 8 in the ones' place.
Because 8 is an even number,
7138 is divisible by 2.

Check: $2\overline{)7138}$ = 3569 R0

Use the divisibility rule for 2 to decide if each number is divisible by 2. Answer *yes* or *no*.

1. 34   **2.** 43   **3.** 317   **4.** 318   **5.** 801

6. 910   **7.** 2439   **8.** 6025   **9.** 15,432   **10.** 27,156

The divisibility rule for 2 simply requires looking at the ones' place. This is also true of the divisibility rules for 5 and 10.

Example 2

Renee used mental math to decide whether or not each number below is divisible by 5 and 10. What were the results?

Number:	45	67	950	2315	39,190
Divisible by 5?					
Divisible by 10?					

■ Solution ■

Number:	45	67	950	2315	39,190
Divisible by 5?	yes	no	yes	yes	yes
Divisible by 10?	no	no	yes	no	yes

(continued)

Bridge Unit, PASSPORT TO MATHEMATICS BOOK 1

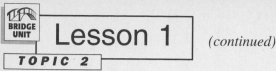
11. What patterns do you see in the completed table?

12. Complete the sentence, "A number is divisible by 5 if…".

13. Complete the sentence, "A number is divisible by 10 if…".

Some tests for divisibility involve working with the entire number.

Example 3

How can Evan decide whether or not 4698 is divisible by 3?

How can he decide whether or not 4698 is divisible by 9?

■ Solution ■

A number is divisible by 3 if the sum of its digits is divisible by 3.

$4 + 6 + 9 + 8 = 27$

$27 \div 3 = 9 \text{ R0}$

So, 4698 is divisible by 3.

A number is divisible by 9 if the sum of its digits is divisible by 9.

$4 + 6 + 9 + 8 = 27$

$27 \div 9 = 3 \text{ R0}$

So, 4698 is divisible by 9.

Using the divisibility rules for 3 and 9, decide whether or not each number is divisible by 3, by 9, by both 3 and 9, or by neither.

14. 345   **15.** 774   **16.** 969   **17.** 210   **18.** 889

19. 4536   **20.** 9810   **21.** 6207   **22.** 15,589   **23.** 34,216

24. How are the divisibility rules for 3 and 9 alike?

Spiral Review

25. Write three numbers that round to 4000 when rounded to the nearest thousand.

26. A tailor wants to cut 9.5 cm lengths of ribbon from a 1 meter bolt. How many lengths can the tailor cut?

Warm-Ups

FOR USE WITH TOPIC 2, LESSON 1

Standardized Testing Warm-Ups

For Exercises 1–4, use the table below. It shows T-shirt prices and Saturday sales at a T-shirt shop.

1. Estimate the total sales of plain T-shirts.

   A about $50

   B about $60

   C about $100

   D about $120

Shirt type	Cost	Items sold
plain T-shirt	$5.99	19
shirt with decal	$7.79	22
shirt with printing	$8.59	13
shirt with decal and printing	$10.19	9

2. Which type of T-shirt brought in the most income?

   A plain   **B** decal   **C** printing   **D** decal and printing

Homework Review Warm-Ups

Use the table above.

3. How many decal T-shirts can you buy with $50?

4. You have $35 to spend. Give two examples of what you could buy without going over your budget.

Project

FOR USE WITH TOPIC 2, LESSON 1

In Topic 1, Lesson 4, you completed an order list for the school store items you chose. Look back at the table to help you recall the items. Then complete the table below to show whether or not you can divide and sell each item in groups of 2, 3, 5, 9, or 10 with none left over.

Item	Number ordered	Can divide into groups of:
	28	
	36	
	42	
	50	
	60	

Bridge Unit, PASSPORT TO MATHEMATICS BOOK 1

BRIDGE UNIT
Activity 1

FOR USE WITH TOPIC 2, LESSON 1

Finding Divisibility Patterns

In this activity, you will look more closely at divisibility patterns. Then you will write a new divisibility rule of your own!

What to do:

1. Use the list of numbers below.

21	22	24	25	90
111	119	130	504	621
1140	1203	1377	1757	2005
18,241	18,242	35,469	76,540	82,621

Use divisibility rules to decide whether or not each number is divisible by 2, 3, 5, 9, and 10. Then list all the numbers that belong in each category below.

Divisible by 2:

Divisible by 3:

Divisible by 5:

Divisible by 9:

Divisible by 10:

2. a. Are there numbers that are divisible by 3 that are *not* divisible by 9?

b. Are there numbers that are divisible by 9 that are *not* divisible by 3?

3. Find the numbers that are divisible by both 2 *and* 5. Are they on the "Divisible by 10" list? How can you explain this?

4. a. Find the numbers that are divisible by both 2 *and* 3. Divide each by 6. (You can use a calculator to help.)

b. What did you find in part (a)? How can you explain this?

c. Write a divisibility rule for 6.

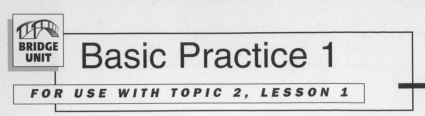

Basic Practice 1

Complete the table. Answer *yes* or *no* for each box.

Number	Divisible by 2	Divisible by 3	Divisible by 5	Divisible by 9	Divisible by 10
1. 324					
2. 475					
3. 525					
4. 600					
5. 1234					
6. 3951					
7. 4230					
8. 7803					
9. 9360					
10. 11,235					
11. 15,972					
12. 23,409					

13. Write three numbers that are divisible by the given number. Do not repeat numbers from Exercises 1–12.

    a. 2

    b. 3

    c. 5

    d. 6

    e. 9

    f. 10

14. A giant pizza is divided into 18 pieces. What are the different numbers of people you can divide it among so that there are no pieces left over?

15. You and 8 of your friends have been saving money by recycling aluminum cans. You have made a total of $58.86. Can you divide the money evenly among yourselves? How can you tell without dividing?

16. What is the smallest number you can find that is divisible by 2, 3, 5, 6, 9, and 10? Explain how you can use the divisibility rules to find the number.

Bridge Unit, PASSPORT TO MATHEMATICS BOOK 1

Goal
List factors of numbers. Use factor lists to identify the common factors and greatest common factor of two or more numbers.

Greatest Common Factor

Dominique has 8 red balloons and 12 blue balloons to decorate tables for a party. At each table where she puts balloons, she wants the same numbers of each color. Dominique can use common factors to find how many tables to decorate. She can use the greatest common factor to find the most tables she can decorate this way.

Terms to Know	*Example / Illustration*
Factor a whole number that divides another whole number with a remainder of 0	The factors of 8 are 1, 2, 4, and 8. The factors of 12 are 1, 2, 3, 4, 6, and 12.
Common factor a number that is a factor of two or more numbers	The common factors of 8 and 12 are 1, 2, and 4.
Greatest common factor the largest common factor of two or more numbers	The greatest common factor of 8 and 12 is 4. Dominique can decorate 4 tables with the balloons by putting 2 red balloons and 3 blue balloons at each of the tables.

UNDERSTANDING THE MAIN IDEAS

To find the greatest common factor, or GCF, of two numbers, first list the factors of each number. The GCF is the largest number that is on both lists.

(continued)

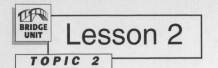

Lesson 2 *(continued)*

Example 1

Find the greatest common factor (GCF) of 12 and 16.

■ Solution ■

Step 1: List the factors of each number in order from least to greatest.

Factors of 12: 1, 2, 3, 4, 6, 12

Factors of 16: 1, 2, 4, 8, 16

Step 2: Identify the factors common to both lists.

Common factors of 12 and 16: 1, 2, 4

Step 3: The largest of the common factors is 4. This is the GCF.

The GCF of 12 and 16 is 4.

List the factors of each number. Circle or underline the common factors. Then write the greatest common factor (GCF).

1. 8, 20   **2.** 9, 24   **3.** 10, 15   **4.** 11, 14

5. 4, 10   **6.** 12, 24   **7.** 35, 42   **8.** 30, 45

(continued)

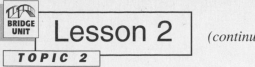
In a similar way, you can also find the greatest common factor of three or more numbers.

Example 2

Find the greatest common factor (GCF) of 6, 15, and 21.

Solution

Step 1: List the factors of each number.

Factors of 6: 1, 2, 3, 6

Factors of 15: 1, 3, 5, 15

Factors of 21: 1, 3, 7, 21

Step 2: Identify the factors common to both lists.

Common factors of 6, 15, and 21: 1, 3

Step 3: The largest of the common factors is 3. This is the GCF.

The GCF of 6, 15, and 21 is 3.

Find the GCF of each group of numbers.

9. 12, 15, 21   **10.** 10, 25, 40   **11.** 8, 18, 28   **12.** 13, 16, 20

13. 7, 14, 28   **14.** 6, 9, 18   **15.** 14, 18, 22   **16.** 12, 24, 30

....
Spiral Review

17. Write three numbers that are each divisible by 2, 4, and 5.

18. A bottle holds 250 milliliters. Can you pour a liter of water into the bottle? Explain.

Warm-Ups

FOR USE WITH TOPIC 2, LESSON 2

Standardized Testing Warm-Ups

1. Which number is *not* a factor of 8, 12, and 24?

    A 1   **B** 2   **C** 3   **D** 4

2. Which list gives all the factors of 36?

    A 1, 2, 3, 6, 8, 12, 18, 36

    B 1, 2, 3, 4, 6, 9, 12, 18, 36

    C 2, 3, 4, 6, 9, 12, 18

    D 1, 2, 3, 4, 6, 8, 12, 36

Homework Review Warm-Ups

Determine whether or not each number is divisible by 2, 3, 5, 9, or 10.

3. 35   **4.** 87   **5.** 105   **6.** 49   **7.** 90

8. What digit can you fill in the blank with to make 47_?_3 divisible by 9?

Project

FOR USE WITH TOPIC 2, LESSON 2

A bulletin board outside the school store advertises the items for sale. You want to cover the bulletin board with attractive backing paper. You decide to cut construction paper squares of two alternating colors so that they fit the bulletin board exactly.

    1. What is the largest size paper square that will fit the length and width of the bulletin board exactly?

    2. How many of these paper squares will fit in each row of the bulletin board? in each column?

    3. How many paper squares of each color do you need?

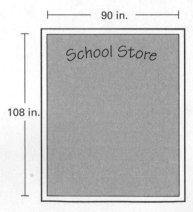

90 in.

108 in.

School Store

Activity 2

MANIPULATIVES

Cooperative Learning

Greatest Common Factors with Rectangles

In this activity, you will find greatest common factors by forming rectangles with unit squares (squares that are one unit on each side). The rectangles that you can make with 6 unit squares are shown below.

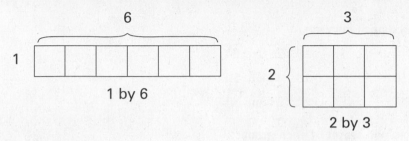

1 by 6   2 by 3

Notice that the height and width of each rectangle are factors of 6.

You will need:

• grid paper

What to do:

1. **a.** Work with a partner. Draw all the rectangles that use 12 unit squares. You don't need to draw both the horizontal and vertical versions of rectangles that are the same size, like 2 by 6 and 6 by 2. Label the lengths and widths.

   b. Write a multiplication fact to describe each rectangle in part (a).

   c. Record the factors from the multiplication facts in a list from least to greatest.

2. Repeat the process for rectangles that use 15 unit squares.

3. Repeat the process for rectangles that use 18 unit squares.

After completing the activity, write answers to the following questions:

4. What factors are on all three lists?

5. Look at the rectangles you drew. Can you see the factors that appear on all three lists? Explain.

6. What is the greatest common factor on all three lists?

7. Describe how you could use rectangles made of unit squares to find the greatest common factor of 9, 15, and 24.

BRIDGE UNIT

Basic Practice 2

Write the factors of each number from least to greatest.

1. 6   **2.** 10   **3.** 12   **4.** 15

5. 18   **6.** 21   **7.** 24   **8.** 30

9. 32   **10.** 36   **11.** 48   **12.** 50

Find the greatest common factor of each pair of numbers.

13. 6 and 10   **14.** 12 and 18   **15.** 21 and 24

16. 30 and 36   **17.** 24 and 36   **18.** 48 and 50

19. 18 and 24   **20.** 32 and 48   **21.** 24 and 50

In Exercises 22–30, find the greatest common factor of each group of numbers.

22. 6, 10, 12   **23.** 10, 15, 30   **24.** 18, 24, 32

25. 12, 24, 36   **26.** 6, 12, 21   **27.** 24, 36, 50

28. 12, 21, 24   **29.** 21, 30, 36   **30.** 18, 36, 48

31. Two different classrooms contain 24 and 32 students. The teachers want to divide each classroom into groups so that all of the groups are the same size. For every student to be in a group, what is the largest possible group size?

32. Give two pairs of numbers whose greatest common factor is 1. Do not choose numbers that differ only by 1 (like 2 and 3 or 5 and 6).

33. What common factors do any two even whole numbers share?

Goal
Identify numbers as prime or composite. Find the prime factorization of a number.

Prime and Composite Numbers

Michael has baked a pan of oatmeal cookies. He cooked 19 cookies on the pan. Without breaking them, the only way for him to divide the cookies evenly among any number of friends is to give one cookie to each of 19 people. This is because 19 is a prime number.

Terms to Know	**Example / Illustration**
Prime number a whole number greater than 1 that has exactly two factors, itself and 1	The first six prime numbers are 2, 3, 5, 7, 11, and 13.
Composite number a whole number that has three or more factors.	Some composite numbers are 4, 15, 24, and 60.
Prime factorization a number written as the product of its prime factors	The prime factorization of 24 is $2 \times 2 \times 2 \times 3$.

UNDERSTANDING THE MAIN IDEAS

A whole number greater than 1 must be either prime or composite. You can use pictures to see how prime and composite numbers differ. Picture a number as a rectangle made of *unit squares* (squares one unit on each side). Two ways to picture the number 8 are shown below. The height and width of each rectangle are factors of 8.

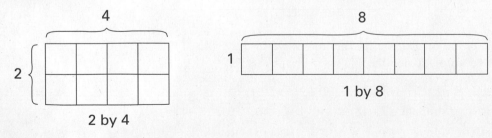

2 by 4

1 by 8

(continued)

Example 1

Decide if each number is prime or composite. Represent each number with rectangles to help you.

   a. 5   **b.** 9

■ Solution ■

Step 1: Draw all of the possible sizes of rectangle for each number. You do not need to draw both horizontal and vertical rectangles that are actually the same size.

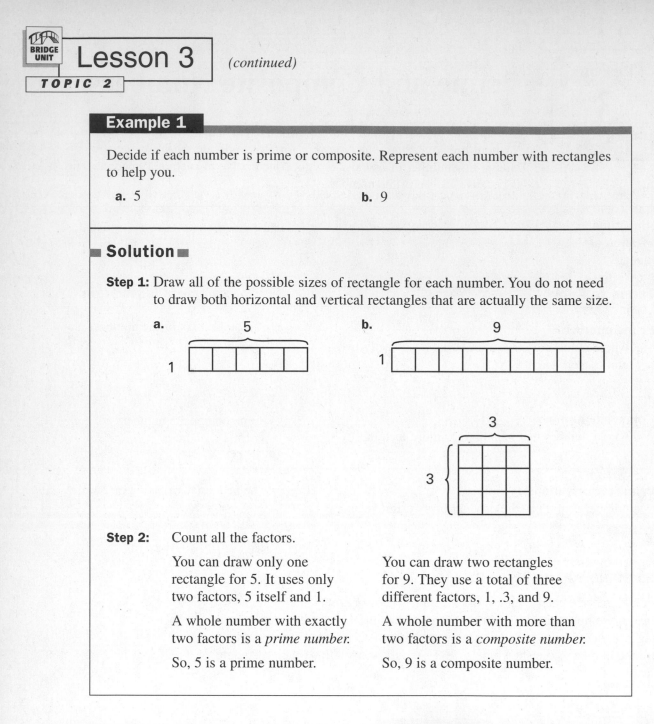

Step 2:   Count all the factors.

You can draw only one rectangle for 5. It uses only two factors, 5 itself and 1.	You can draw two rectangles for 9. They use a total of three different factors, 1, .3, and 9.
A whole number with exactly two factors is a *prime number.*	A whole number with more than two factors is a *composite number.*
So, 5 is a prime number.	So, 9 is a composite number.

Determine if each number is *prime* or *composite*. You may want to draw the rectangles for each number to help you decide.

1. 6   **2.** 11   **3.** 13   **4.** 14

5. 21   **6.** 17   **7.** 18   **8.** 19

9. 23   **10.** 25   **11.** 27   **12.** 29

(continued)

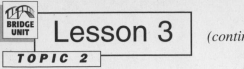

Lesson 3 *(continued)*

You can write any composite number as the product of its prime factors.
To find the prime factorization of a number, use a *factor tree*.

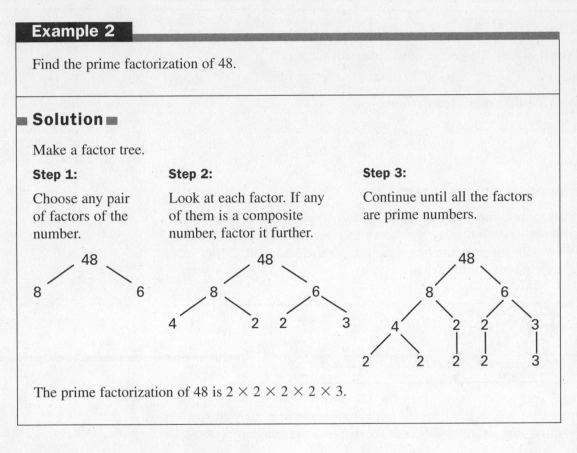

Example 2

Find the prime factorization of 48.

Solution

Make a factor tree.

Step 1:
Choose any pair of factors of the number.

Step 2:
Look at each factor. If any of them is a composite number, factor it further.

Step 3:
Continue until all the factors are prime numbers.

The prime factorization of 48 is $2 \times 2 \times 2 \times 2 \times 3$.

Write the prime factorization. Use a factor tree.

13. 12   **14.** 18   **15.** 24   **16.** 32

17. 50   **18.** 75   **19.** 98   **20.** 100

. . . .
Spiral Review

21. Find the greatest common factor of 12 and 27.

22. Which of the numbers 2, 3, 5, 6, 9, and 10 is 3672 divisible by?

BRIDGE UNIT

Warm-Ups

Standardized Testing Warm-Ups

1. Which of the following is *not* a composite number?

    A 15   **B** 27   **C** 21   **D** 23

2. Which is the prime factorization of 36?

    A $2 \times 2 \times 2 \times 2 \times 2$ **B** $2 \times 2 \times 3 \times 3$

    C 6×6 **D** $2 \times 2 \times 9$

Homework Review Warm-Ups

3. A toy store owner has 6 toy lions, 12 tigers, and 15 bears to use for displays. At each display, she wants the same numbers of each animal. What is the greatest number of displays that she can make? How many of each animal are in each display?

Project

You are organizing your school store. You have room to display 36 small boxed staplers. You want to arrange them in the shape of a rectangle.

1. Draw all the rectangles you can from 36 unit squares to represent your display. You do not need to draw both the horizontal and vertical versions of rectangles that are the same size.

2. For each rectangle except the 1-by-36 rectangle, write the prime factorization of 36. Then choose two groups (one of these may be a single factor) of the factors to represent that rectangle. For example, for the 6-by-6 rectangle, you can divide the prime factors into the two groups 2×3 and 2×3.

3. Can you make any groupings of the prime factors that give you a new size rectangle? If you can, draw the rectangle.

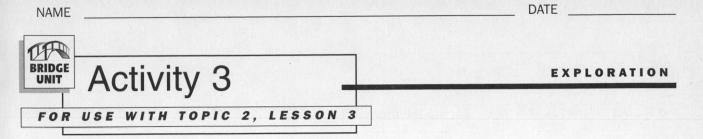

Activity 3

EXPLORATION

FOR USE WITH TOPIC 2, LESSON 3

Sieve of Eratosthenes

The ancient Greek mathematician Eratosthenes invented a method for finding prime numbers. Use his method to find the primes for numbers through 100.

What to do:

- Use the number table below. Cross out the number 1. It is neither prime nor composite.

- Circle 2, since it is prime. Cross out all multiples of 2 (4, 6, 8, ... 98, 100).

- Circle 3, 5, and 7, since they are prime. Cross out all multiples of each number.

- Circle the remaining numbers.

1	2	3	4	5	6	7	8	9	10
11	12	13	14	15	16	17	18	19	20
21	22	23	24	25	26	27	28	29	30
31	32	33	34	35	36	37	38	39	40
41	42	43	44	45	46	47	48	49	50
51	52	53	54	55	56	57	58	59	60
61	62	63	64	65	66	67	68	69	70
71	72	73	74	75	76	77	78	79	80
81	82	83	84	85	86	87	88	89	90
91	92	93	94	95	96	97	98	99	100

After completing the activity, write answers to the following questions.

1. How do you know that 2, 3, 5, and 7 are prime numbers?

2. Are the crossed-out numbers prime or composite? Tell how you know.

3. Are the circled numbers prime or composite? Tell how you know.

BRIDGE UNIT

Basic Practice 3

FOR USE WITH TOPIC 2, LESSON 3

In Exercises 1–24, determine if each number is *prime* or *composite.* You may want to draw the rectangles for each number to help you decide.

1. 3	**2.** 4	**3.** 8	**4.** 10
5. 12	**6.** 19	**7.** 31	**8.** 37
9. 39	**10.** 41	**11.** 43	**12.** 49
13. 53	**14.** 57	**15.** 61	**16.** 63
17. 73	**18.** 78	**19.** 81	**20.** 83
21. 87	**22.** 93	**23.** 94	**24.** 97

25. Is it true that *every* even number greater than two is a composite number? Give a reason for your answer.

26. Can a number be both prime *and* composite? Explain.

Write the prime factorization. Use a factor tree.

27. 20	**28.** 27	**29.** 54	**30.** 63
31. 64	**32.** 70	**33.** 104	**34.** 99

In Exercises 35–37, find the number represented by each prime factorization.

35. $2 \times 2 \times 3 \times 5$ **36.** $3 \times 3 \times 3 \times 3$ **37.** $2 \times 2 \times 3 \times 7$

38. Complete the prime factorization for each factor tree below.

What happens? How can you explain this?

Bridge Unit, PASSPORT TO MATHEMATICS BOOK 1

Basic Assessment

BRIDGE UNIT

FOR USE WITH TOPIC 2

Complete the table. Answer *yes* or *no* for each box.

Number	Divisible by 2	Divisible by 3	Divisible by 5	Divisible by 9	Divisible by 10
1. 87					
2. 312					
3. 865					
4. 840					
5. 3861					

Complete each number with a digit so that the number is divisible by the given factors.

6. 22?; divisible by 3 and 5

7. 41?; divisible by 2 and 3

8. 6?0; divisible by 9 and 10

9. ?98; divisible by 2 and 9

Find the greatest common factor of each pair or group of numbers.

10. 5 and 20

11. 12 and 15

12. 6 and 14

13. 8 and 18

14. 21 and 28

15. 9 and 19

16. 3, 9, and 21

17. 6, 12, and 15

18. 10, 15, and 30

19. 11, 15, and 20

20. 15, 30, and 45

21. 14, 28, and 35

In Exercises 22–29, identify each number as *prime* or *composite*. If the number is composite, find its prime factorization.

22. 21

23. 35

24. 47

25. 56

26. 61

27. 72

28. 85

29. 91

30. What is the greatest common factor of any two prime numbers? How do you know?

Answers

Lesson 1

Lesson Exercises *pages 52–53*

1. Yes. **2.** No. **3.** No.

4. Yes. **5.** No. **6.** Yes.

7. No. **8.** No. **9.** Yes.

10. Yes.

11. Answers will vary. Possible answer: Some numbers were divisible by only 5, others by 5 and 10. No numbers were divisible by only 10.

12. A number is divisible by 5 if the digit in the ones' place is a 5 or a 0.

13. A number is divisible by 10 if the digit in the ones' place is zero.

14. 3 only **15.** both **16.** 3 only

17. 3 only **18.** neither **19.** both

20. both **21.** 3 only **22.** neither

23. neither

24. The divisibility rules for 3 and 9 both involve checking the divisibility of the sum of the digits of the number.

25. Answers will vary, but must be between 3500 and 4499.

26. ten lengths

Warm-Ups *page 54*

1. D **2.** B **3.** 6

4. One possible answer: One of each type of T-shirt.

Project *page 54*

28: 2;
36: 2, 3, 9;
42: 2, 3;
50: 2, 5, 10;
60: 2, 3, 5, 10;

Activity 1 *page 55*

1. Divisible by 2: 22; 24; 90; 130; 504; 1140; 18,242; 76,540

Divisible by 3: 21; 24 90; 111; 504; 621; 1140; 1203; 1377; 35,469

Divisible by 5: 25; 90; 130; 1140; 2005; 76,540

Divisible by 9: 90; 504; 621; 1377; 35,469

Divisible by 10: 90; 130; 1140; 76,540

2. a. Yes. **b.** No.

3. 90; 130; 1140; 76,540; Yes.; If a number is divisible 2 it ends in 0, 2, 4, 6, or 8, and if it is divisible by 5 it ends in 0 or 5. The only number that is on both lists is 0, and a number that ends in 0 is divisible by 10.

4. a. 24, 90, 504, 1140; 4, 15, 84, 190

b. Each number is divisible by 6. If a number is divisible by 2 and 3, it has both 2 and 3 as factors, so it is divisible by $2 \times 3 = 6$.

c. If a number is divisible by both 2 and 3, then it is divisible by 6.

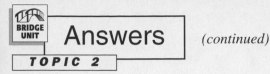

Basic Practice 1 *page 56*

Number	Divisible by 2	Divisible by 3	Divisible by 5	Divisible by 9	Divisible by 10
1. 324	yes	yes	no	yes	no
2. 475	no	no	yes	no	no
3. 525	no	yes	yes	no	no
4. 600	yes	yes	yes	no	yes
5. 1234	yes	no	no	no	no
6. 3951	no	yes	no	yes	no
7. 4230	yes	yes	yes	yes	yes
8. 7803	no	yes	no	yes	no
9. 9360	yes	yes	yes	yes	yes
10. 11,235	no	yes	yes	no	no
11. 15,972	yes	yes	no	no	no
12. 23,409	no	yes	no	yes	no

13. Answers will vary.

14. 2, 3, 6, 9, or 18

15. Yes. The sum of the digits is 27, which is divisible by 9.

16. Answers may vary. The smallest such number is 90. Sample explanation: You know the number must end in 0 and that the sum of the digits must be divisible by 9. The smallest number like this is 90. Because 90 is divisible by 9, it is divisible by 3, and because it is divisible by 10, it is divisible by 2 and 5. Because it is divisible by 2 and 3, it is also divisible by 6.

Lesson 2

Lesson Exercises *pages 58–59*

1. 8: 1, 2, 4, 8; 20: 1, 2, 4, 5, 10, 20; GCF: 4

2. 9: 1, 3, 9; 24: 1, 2, 3, 4, 6, 8, 12, 24; GCF: 3

3. 10: 1, 2, 5, 10; 15: 1, 3, 5, 15; GCF: 5

4. 11: 1, 11; 14: 1, 2, 7, 14; GCF: 1

5. 4: 1, 2, 4; 10: 1, 2, 5, 10; GCF: 2

6. 12: 1, 2, 3, 4, 6, 12; 24: 1, 2, 3, 4, 6, 12, 24; GCF: 12

7. 35: 1, 5, 7, 35; 42: 1, 2, 3, 6, 7, 14, 21, 42; GCF: 7

8. 30: 1, 2, 3, 5, 6, 10, 15, 30; 45: 1, 3, 5, 15, 45; GCF: 15

9. 3   **10.** 5   **11.** 2

12. 1   **13.** 7   **14.** 3

15. 2   **16.** 6

17. Answers will vary but must be 20 or multiples of 20.

18. No; a liter of water is 1000 milliliters.

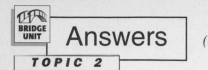

Warm-Ups *page 60*

1. C **2.** B **3.** 5

4. 3 **5.** 3, 5 **6.** none

7. 2, 3, 5, 9, 10 **8.** 4

Project *page 60*

1. 18 inches

2. 5 squares in each row; 6 squares in each column

3. 15 squares of each color

Activity 2 *page 61*

1. a. Check students' rectangles. Rectangles should be 1 by 12, 2 by 6, and 3 by 4.

  b. $1 \times 12 = 12$; $2 \times 6 = 12$; $3 \times 4 = 12$

  c. 1, 2, 3, 4, 6, 12

2. Rectangles should be 1 by 15 and 3 by 5.; $1 \times 15 = 15$, $3 \times 5 = 15$; 1, 3, 5, 15

3. Rectangles should be 1 by 18, 2 by 9, and 3 by 6.; $1 \times 18 = 18$, $2 \times 9 = 18$, $3 \times 6 = 18$; 1, 2, 3, 6, 9, 18

4. 1 and 3.

5. Yes; one rectangle from all three groups consists of 3 rows (or columns); one rectangle from all three groups consists of 1 row.

6. 3

7. Draw all possible rectangles for each number. Make a list showing the factors that form the rectangles. Find the greatest factor common to all three lists.

Basic Practice 2 *page 62*

1. 1, 2, 3, 6 **2.** 1, 2, 5, 10

3. 1, 2, 3, 4, 6, 12 **4.** 1, 3, 5, 15

5. 1, 2, 3, 6, 9, 18 **6.** 1, 3, 7, 21

7. 1, 2, 3, 4, 6, 8, 12, 24

8. 1, 2, 3, 5, 6, 10, 15, 30

9. 1, 2, 4, 8, 16, 32

10. 1, 2, 3, 4, 6, 9, 12, 18, 36

11. 1, 2, 3, 4, 6, 8, 12, 16, 24, 48

12. 1, 2, 5, 10, 25, 50

13. 2 **14.** 6 **15.** 3

16. 6 **17.** 12 **18.** 2

19. 6 **20.** 16 **21.** 2

22. 2 **23.** 5 **24.** 2

25. 12 **26.** 3 **27.** 2

28. 3 **29.** 3 **30.** 6

31. 8 students

32. Answers will vary. Samples: 4 and 9 or 6 and 13

33. 1 and 2

Lesson 3

Lesson Exercises *pages 64–65*

1. composite **2.** prime

3. prime **4.** composite

5. composite **6.** prime

7. composite **8.** prime

9. prime **10.** composite

11. composite **12.** prime

13. $2 \times 2 \times 3$ **14.** $2 \times 3 \times 3$

15. $2 \times 2 \times 2 \times 3$ **16.** $2 \times 2 \times 2 \times 2 \times 2$

17. $2 \times 5 \times 5$ **18.** $3 \times 5 \times 5$

19. $2 \times 7 \times 7$ **20.** $2 \times 2 \times 5 \times 5$

21. 3 **22.** 2, 3, 6, and 9

Warm-Ups *page 66*

1. D **2.** B

3. 3 displays; 2 lions, 4 tigers, and 5 bears

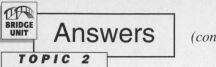
Project *page 66*

1. Check students' answers. Rectangles should be 1 by 36, 2 by 18, 3 by 12, 4 by 9, and 6 by 6.

2. 2 by 18: 2 and $2 \times 3 \times 3$; 3 by 12: 3 and $2 \times 2 \times 3$; 4 by 9: 2×2 and 3×3; 6 by 6: 2×3 and 2×3

3. Answers will vary. If students have drawn all the rectangles in Question 1, they will not be able to make any new groupings of the prime factors.

Activity 3 *page 67*

1. Their only factors are themselves and 1.

2. The crossed-out numbers are composite. Since they are multiples of 2, 3, 5, or 7, they have factors other than themselves and 1.

3. The circled numbers are prime. They are not multiples of 2, 3, 5, or 7. Their only factors are themselves and 1.

Basic Practice 3 *page 68*

1. prime
2. composite
3. composite
4. composite
5. composite
6. prime
7. prime
8. prime
9. composite
10. prime
11. prime
12. composite
13. prime
14. composite
15. prime
16. composite
17. prime
18. composite
19. composite
20. prime
21. composite
22. composite
23. composite
24. prime

25. Yes. Possible explanation: Every even number greater than 2 is a multiple of 2. A prime number has only itself and 1 as factors. If 2 is also a factor, the number has more than two factors and therefore must be composite.

26. No. Possible explanation: If a number is prime, it has exactly two factors. If it is composite, it has more than two factors. A number cannot have exactly two factors and more than two factors at the same time.

27. $2 \times 2 \times 5$
28. $3 \times 3 \times 3$
29. $2 \times 3 \times 3 \times 3$
30. $3 \times 3 \times 7$
31. $2 \times 2 \times 2 \times 2 \times 2 \times 2$
32. $2 \times 5 \times 7$
33. $2 \times 2 \times 2 \times 13$
34. $3 \times 3 \times 11$
35. 60
36. 81
37. 84
38. a. $2 \times 2 \times 2 \times 3$

   b. $3 \times 2 \times 2 \times 2$

   c. In each case, the prime factorization is $2 \times 2 \times 2 \times 3$. There is only one prime factorization for any number; the different trees just give a different *arrangement* of the factors.

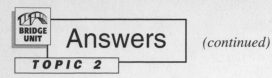

Topic 2 Basic Assessment *page 69*

Number	Divisible by 2	Divisible by 3	Divisible by 5	Divisible by 9	Divisible by 10
1. 87	no	yes	no	no	no
2. 312	yes	yes	no	no	no
3. 865	no	no	yes	no	no
4. 840	yes	yes	yes	no	yes
5. 3861	no	yes	no	yes	no

6. 5   **7.** 4   **8.** 3

9. 1   **10.** 5   **11.** 3

12. 2   **13.** 2   **14.** 7

15. 1   **16.** 3   **17.** 3

18. 5   **19.** 1   **20.** 15

21. 7

22. composite; 3×7

23. composite; 5×7

24. prime

25. composite; $2 \times 2 \times 2 \times 7$

26. prime

27. composite; $2 \times 2 \times 2 \times 3 \times 3$

28. composite; 5×17

29. prime

30. 1; Any prime number has only itself and 1 as factors. So, if the prime numbers are different, they can share only the factor 1.

Lesson

1

Goal

Determine when a fraction is in simplest form. Use the GCF to find the simplest form of a fraction.

Fractions in Simplest Form

One day during the winter cold and flu season, 56 of the 336 students in Grant's grade were out sick. You can write this as the fraction $\frac{56}{336}$. But the fraction would probably mean a lot more to you if you saw it in its simpler equivalent form of $\frac{1}{6}$.

Terms to Know	**Example / Illustration**
Equivalent fractions fractions that represent the same number	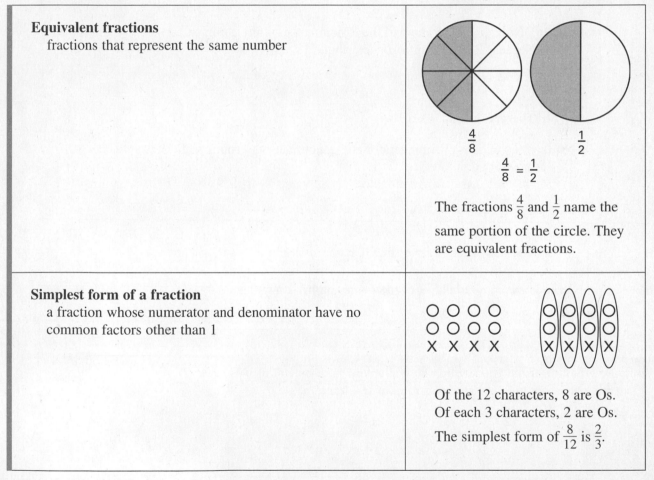
Simplest form of a fraction a fraction whose numerator and denominator have no common factors other than 1	

UNDERSTANDING THE MAIN IDEAS

As long as you can divide the numerator and denominator of a fraction by common factors, you can simplify the fraction.

(continued)

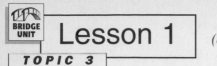

Lesson 1 *(continued)*

Example 1

Nicole has 24 games on her computer. She already knows how to play 18 of them. What is the simplest form of the fraction that describes the portion of computer games that Nicole knows how to play?

■ Solution ■

Step 1: Write the fraction that represents the situation.

The fraction is $\frac{18}{24}$.

Step 2: Divide the numerator and the denominator by any common factor. Because 2 is a common factor, divide 18 and 24 by 2.

$$\frac{18 \div 2}{24 \div 2} = \frac{9}{12}$$

The new fraction is $\frac{9}{12}$.

Step 3: Keep dividing the numerator and denominator by common factors until the greatest common factor is 1.
Because 3 is a common factor, divide 9 and 12 by 3.

$$\frac{9 \div 3}{12 \div 3} = \frac{3}{4}$$

The new fraction is $\frac{3}{4}$. The GCF of 3 and 4 is 1.

Nicole knows how to play $\frac{3}{4}$ of the games on her computer.

Indicate whether or not the fraction is in simplest form by writing *yes* or *no*. If the answer is *no*, give the simplest form of the fraction.

1. $\frac{3}{6}$ 2. $\frac{6}{8}$ 3. $\frac{12}{20}$ 4. $\frac{17}{25}$

5. $\frac{8}{12}$ 6. $\frac{4}{10}$ 7. $\frac{5}{12}$ 8. $\frac{50}{100}$

Another way to find the simplest form of a fraction is first to find the greatest common factor of the numerator and the denominator. Then you can divide the numerator and denominator by the greatest common factor.

(continued)

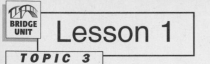
Example 2

Neeti made a patchwork quilt out of fabric scraps. She had 30 fabric scraps. Of these, 18 were prints or stripes. What is the simplest form of the fraction that describes the portion of the scraps that were prints or stripes?

■ Solution ■

Step 1: Write the fraction that represents the situation.

The fraction is $\frac{18}{30}$.

Step 2: Find the greatest common factor of the numerator and denominator of the fraction.

Factors of 18: 1, 2, 3, 6, 9, 18

Factors of 30: 1, 2, 3, 5, 6, 10, 15, 30

Common factors of 18 and 30: 1, 2, 3, 6

The GCF of 18 and 30 is 6.

Step 3: Divide the numerator and denominator by the greatest common factor.

$$\frac{18 \div 6}{30 \div 6} = \frac{3}{5}$$

The portion of Neeti's fabric scraps that were prints or stripes is $\frac{3}{5}$.

Use the greatest common factor to write each fraction in simplest form. If it is already in simplest form, write *simplest form.*

9. $\frac{12}{28}$ **10.** $\frac{14}{42}$ **11.** $\frac{21}{35}$ **12.** $\frac{15}{45}$

13. $\frac{24}{36}$ **14.** $\frac{18}{48}$ **15.** $\frac{35}{40}$ **16.** $\frac{23}{34}$

Spiral Review

17. By which of the numbers 2, 3, 5, 9, and 10 is 5412 divisible?

18. Complete the statement: The _____ of 76 is $2 \times 2 \times 19$.

Warm-Ups

BRIDGE UNIT

FOR USE WITH TOPIC 3, LESSON 1

Standardized Testing Warm-Ups

1. Which of the following fractions is in simplest form?

A $\frac{17}{51}$ **B** $\frac{11}{33}$ **C** $\frac{23}{26}$ **D** $\frac{12}{27}$

2. Which one of the following fractions is not equivalent to the others?

A $\frac{5}{6}$ **B** $\frac{21}{24}$ **C** $\frac{15}{18}$ **D** $\frac{35}{42}$

Homework Review Warm-Ups

3. List the prime numbers between 40 and 50.

4. Use a factor tree to find the prime factorization of 90.

Project

FOR USE WITH TOPIC 3, LESSON 1

In Topic 1, Lesson 5, you began to think about how long each student would need to work in the school store each month. Next, you will need to think about actually scheduling work times for each student.

1. Suppose you have decided that each student needs to work for 1 hour each month. By talking to students, you find that it's best to schedule them to work from 15 to 45 minutes at a time, counting in 5 minute blocks. Write fractions for the possible portions of their monthly time that students might work at one time.

2. Write the fractions from Question 1 in simplest form.

3. On a particular day, you are scheduling 5 students to work at the store. The store will be open from 11:00 to 1:30 that day. The amounts of time each student will work are shown below.

Renee: 15 min   Lisa: 20 min   LaToya: 30 min
Benjamin: 40 min   Malcolm: 45 min

La Toya can only work from 12:00 to 12:30. Lisa can only work after noon. Complete the work schedule below for the day.

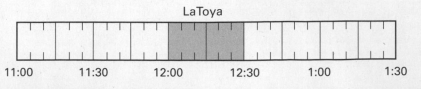

Bridge Unit, PASSPORT TO MATHEMATICS BOOK 1

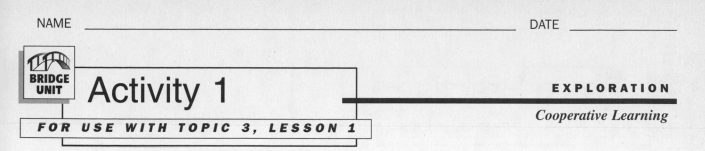

Activity 1

FOR USE WITH TOPIC 3, LESSON 1

Identifying Equivalent Fractions

In this activity, you will assemble pieces of circles to find equivalent fractions and identify the simplest form of a fraction.

You will need:

- paper
- scissors

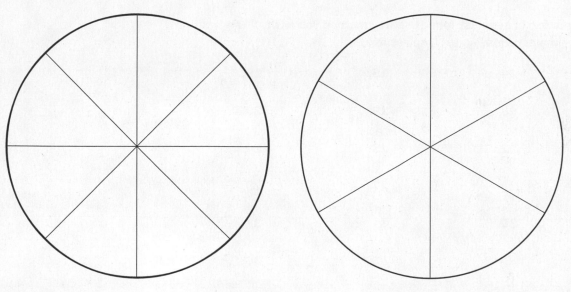

What to do:

1. Work in a group of 3–4 students. Trace and cut out 5 paper circles, using the patterns above. Use the pattern on the left to draw 3 of the circles, one divided into halves, one into fourths, and one into eighths. Use the circle on the right to draw 2 circles, one divided into thirds and one into sixths.

2. Cut each circle into pieces along the lines. Label each piece with the fraction for the part of the circle that it represents.

3. Use the pieces to model each fraction below. Then try to model an equivalent fraction using fewer pieces. Continue until you cannot form an equivalent fraction using a set with fewer pieces.

   a. $\frac{2}{4}$ **b.** $\frac{2}{6}$ **c.** $\frac{2}{8}$ **d.** $\frac{4}{6}$

   e. $\frac{4}{8}$ **f.** $\frac{6}{8}$

4. How can you tell when one fraction is equivalent to another?

5. How do you know when you have found the equivalent fraction in simplest form?

Basic Practice 1

BRIDGE UNIT

FOR USE WITH TOPIC 3, LESSON 1

Indicate whether or not the fraction is in simplest form. Write *yes* or *no*.

1. $\dfrac{5}{15}$ 2. $\dfrac{2}{7}$ 3. $\dfrac{9}{16}$ 4. $\dfrac{8}{32}$

5. $\dfrac{25}{35}$ 6. $\dfrac{10}{13}$ 7. $\dfrac{4}{12}$ 8. $\dfrac{16}{21}$

9. $\dfrac{6}{27}$ 10. $\dfrac{18}{27}$ 11. $\dfrac{19}{20}$ 12. $\dfrac{23}{46}$

Write each fraction in simplest form. Use any method you wish. If the fraction is already in simplest form, write *simplest form*.

13. $\dfrac{16}{27}$ 14. $\dfrac{6}{35}$ 15. $\dfrac{9}{30}$ 16. $\dfrac{4}{16}$

17. $\dfrac{7}{28}$ 18. $\dfrac{40}{70}$ 19. $\dfrac{3}{18}$ 20. $\dfrac{8}{21}$

21. $\dfrac{34}{50}$ 22. $\dfrac{9}{12}$ 23. $\dfrac{21}{35}$ 24. $\dfrac{16}{40}$

25. $\dfrac{12}{25}$ 26. $\dfrac{9}{24}$ 27. $\dfrac{13}{20}$ 28. $\dfrac{6}{16}$

29. $\dfrac{16}{30}$ 30. $\dfrac{50}{75}$ 31. $\dfrac{18}{27}$ 32. $\dfrac{40}{100}$

33. $\dfrac{24}{42}$ 34. $\dfrac{12}{36}$ 35. $\dfrac{18}{73}$ 36. $\dfrac{4}{15}$

Use the table below for Exercises 37–40. Give your answers in simplest form.

The scoreboard shows the wins, losses and ties of the Cougars baseball team.

37. What part of its games did the team win?

38. What part of its games did the team lose?

39. What part of its games did the team tie?

40. What part of its games did the team win or tie?

Cougars team record

Wins	Losses	Ties
16	12	8

Goal
Write decimals as
fractions in simplest
form.

Changing Decimals to Fractions

Gray tiles make up 0.20 of
this 100 square tile design.

The design is $\frac{20}{100}$, or $\frac{1}{5}$, gray.

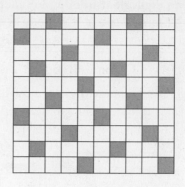

Terms to Know

Decimal
a number with one or more digits to the right of a
decimal point.

Example / Illustration

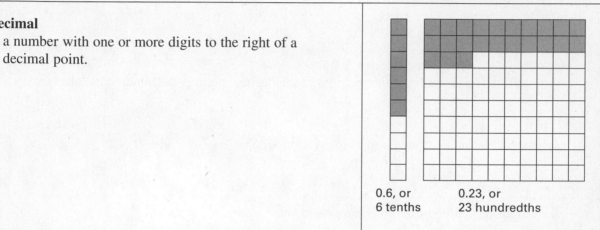

0.6, or
6 tenths

0.23, or
23 hundredths

UNDERSTANDING THE MAIN IDEAS

You can write decimals as fractions in simplest form.

Example 1

Quentin made 0.4 of his free throws in the last basketball game. Write this amount as
a fraction in lowest terms.

◾ Solution ◾

Step 1: Write the decimal as a fraction.
The first place to the right of the
decimal point is the tenths' place.

$$0.4 = \frac{4}{10}$$

Quentin made $\frac{2}{5}$ of his free throws.

Step 2: Write the fraction in simplest form.
The GCF of 4 and 10 is 2.

$$\frac{4 \div 2}{10 \div 2} = \frac{2}{5}$$

(continued)

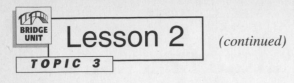

Lesson 2 *(continued)*

TOPIC 3

Write an equivalent fraction for each decimal. Then write the fraction in simplest form if it is not already in simplest form.

1. 0.4   2. 0.1   3. 0.9   4. 0.7

5. 0.3   6. 0.5   7. 0.8   8. 0.2

No matter how many digits follow the decimal point, you can still write the decimal as a fraction in simplest form.

Example 2

Jill's batting average is 0.25. What fraction of the time does Jill make a hit?

■ Solution ■

Step 1: Write the decimal as a fraction.
The second place to the right of the decimal point is the hundredths' place.

$$0.25 = \frac{25}{100}$$

Step 2: Write the fraction in simplest form. The GCF of 25 and 100 is 25.

$$\frac{25 \div 25}{100 \div 25} = \frac{1}{4}$$

Jill makes a hit $\frac{1}{4}$ of the time she gets to bat.

Write an equivalent fraction for each decimal. Then write the fraction in simplest form if it is not already in simplest form.

9. 0.40   10. 0.15   11. 0.28   12. 0.75

13. 0.09   14. 0.18   15. 0.65   16. 0.05

17. 0.12   18. 0.68   19. 0.37   20. 0.01

· · · · · · · · · · · · · · · · · · · ·
Spiral Review

21. Wayne is 1.65 m tall. Sara is 169 cm tall. Who is taller? Explain.

22. Jerry practiced piano for 45 minutes. In simplest form, what fractional part of an hour did he practice?

Warm-Ups

FOR USE WITH TOPIC 3, LESSON 2

Standardized Testing Warm-Ups

1. What is the fraction in simplest form for the decimal 0.72?

A $\frac{36}{50}$ **B** $\frac{7}{10}$ **C** $\frac{18}{25}$ **D** $\frac{19}{25}$

2. Which decimal represents 261 thousandths?

A 2.61   **B** 0.261   **C** 0.0261   **D** 0.00261

Homework Review Warm-Ups

Match each fraction with an equivalent fraction in simplest form.

A $\frac{1}{6}$ **B** $\frac{1}{4}$ **C** $\frac{1}{3}$ **D** $\frac{2}{3}$ **E** $\frac{3}{4}$

3. $\frac{3}{12}$ **4.** $\frac{9}{12}$ **5.** $\frac{4}{12}$ **6.** $\frac{8}{12}$ **7.** $\frac{2}{12}$

Project

FOR USE WITH TOPIC 3, LESSON 2

To help improve your store, you can survey customers. Shalma's class surveys 100 student customers. The survey asks 5 questions about the store. For each question, students can circle "very unsatisfied," "unsatisfied," "somewhat satified," "moderately satisfied," or "very satisfied." Below is Shalma's analysis of the survey information.

- portion who answer "somewhat satisfied" or better to at least 4 questions: 0.75
- portion who answer "somewhat satisfied" or better to all 5 questions: 0.56
- portion who answer "moderately satisfied" or better to all 5 questions: 0.40
- portion who answer "unsatisfied" or worse to more than 1 question: 0.25
- portion who answer "unsatisfied" or worse to all 5 questions: 0.04

1. Write each decimal portion as a fraction in simplest form.

2. Out of 5 students, how many would you expect to answer "moderately satisfied" or better to all 5 questions?

3. Which 2 of the portions together make 1. Can you explain this?

BRIDGE UNIT

Activity 2

FOR USE WITH TOPIC 3, LESSON 2

Decimal Designs

In this activity, you will shade portions of 100 square grids to create "decimal designs." For example, shading 87 squares of the 100 square grid at the right forms the "M" that you see. This is just one of the possible decimal designs for 0.87. Any design or pattern that you can create by shading 87 of the 100 squares will represent 0.87 just as well.

You will need:

- 10 by 10 grid paper

What to do:

Work with a partner.

1. Each partner should pick 4 of the decimals below. Don't tell the other partner your choices. Shade the appropriate part of a 100 square grid to illustrate each decimal of your choice. Make your designs creative.

   a. 0.1   **b.** 0.25   **c.** 0.32   **d.** 0.5

   e. 0.6   **f.** 0.75   **g.** 0.8   **h.** 0.9

2. Write an equivalent fraction in simplest form for each of the decimals above.

3. Trade designs with your partner. Try to guess which of the fractions in simplest form each other's designs represent.

4. Write answers to the following questions.

   a. How did you decide which fraction each design represents?

   b. Which fractions were easiest for you to see from the decimal designs? Why do you think this is so?

   c. Which fractions were most difficult for you to see from the decimal designs? Why do you think this is so?

Basic Practice 2

BRIDGE UNIT

FOR USE WITH TOPIC 3, LESSON 2

Write an equivalent fraction for each decimal. Then write the fraction in simplest form if it is not already in simplest form.

1. 0.6   **2.** 0.60   **3.** 0.02   **4.** 0.04

5. 0.96   **6.** 0.55   **7.** 0.84   **8.** 0.33

For each 100 square grid, give the decimal for the part that is shaded. Write an equivalent fraction for the decimal. Then write the fraction in simplest form if it is not already in simplest form.

9.   **10.**

11.   **12.**

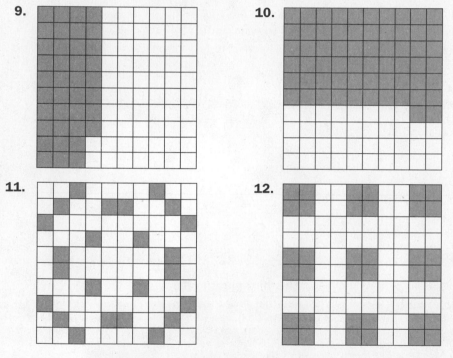

In Exercises 13–14, complete the riddle.

13. I am a decimal between 0.2 and 0.3. I am equivalent to $\frac{1}{4}$. What decimal am I?

14. I am a decimal between 0.01 and 0.1. I am equivalent to $\frac{1}{20}$. What decimal am I?

15. Yvette hits the bull's-eye on her dartboard 0.15 of the time.

     a. Write this decimal as a fraction in simplest form.

     b. If Yvette throws 20 darts, about how many can she predict will hit the bull's-eye?

16. Write the decimals 0.125 and 0.375 as fractions with a denominator of 1000. Then write each fraction in simplest form.

Bridge Unit, PASSPORT TO MATHEMATICS BOOK 1

BRIDGE UNIT

Lesson

3

TOPIC 3

Goal
Rewrite fractions as equivalent fractions with denominators of 10, 100, or 1000. Rewrite fractions as decimals.

Changing Fractions to Decimals

The four great oceans—the Pacific, the Atlantic, the Indian, and the Arctic—alone cover about $\frac{69}{100}$ of Earth's surface. You can rewrite this amount as the decimal 0.69.

UNDERSTANDING THE MAIN IDEAS

You can rewrite a fraction as a decimal number. When the fraction has a denominator of 10, 100, 1000, and so on, it is especially easy to write in decimal form.

Example 1

Write the decimal equivalent for each fraction.

a. $\frac{77}{100}$ b. $\frac{773}{1000}$

■ Solution ■

a. $\frac{77}{100} = 77$ hundredths

The second place to the right of the decimal is the hundredths' place.

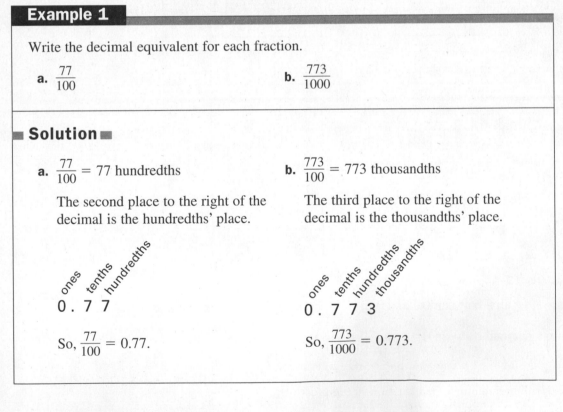

0 . 7 7

So, $\frac{77}{100} = 0.77$.

b. $\frac{773}{100} = 773$ thousandths

The third place to the right of the decimal is the thousandths' place.

0 . 7 7 3

So, $\frac{773}{1000} = 0.773$.

Write the decimal equivalent for each fraction.

1. $\frac{3}{10}$ 2. $\frac{9}{10}$ 3. $\frac{27}{100}$ 4. $\frac{81}{100}$ 5. $\frac{123}{1000}$

(continued)

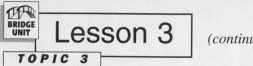
The denominator of $\frac{3}{5}$ is a factor of 10 because $5 \times 2 = 10$. If the denominator of a fraction is a factor of 10, 100, or 1000, you can write an equivalent fraction with a denominator of 10, 100, or 1000. Then it's easy to write the fraction in decimal form.

Example 2

During Amy's vacation, $\frac{3}{4}$ of the days were sunny and warm. Write $\frac{3}{4}$ as a decimal number.

■ Solution ■

Step 1: Notice that 4 is a factor of 100.

$$100 \div 4 = 25$$

Step 2: Multiply to find an equivalent fraction with a denominator of 100.

$$\frac{3 \times 25}{4 \times 25} = \frac{75}{100}$$

Step 3: Rewrite the equivalent fraction as a decimal.

$$\frac{75}{100} = 0.75$$

So, $\frac{3}{4} = \frac{75}{100} = 0.75$. You can check this result by using a calculator to divide 3 by 4.

Rewrite each fraction with a denominator of 10, 100, or 1000. Then write the decimal equivalent.

6. $\frac{2}{5}$ 7. $\frac{1}{2}$ 8. $\frac{47}{50}$ 9. $\frac{2}{25}$ 10. $\frac{8}{50}$

11. $\frac{9}{20}$ 12. $\frac{13}{25}$ 13. $\frac{123}{200}$ 14. $\frac{278}{500}$ 15. $\frac{5}{8}$

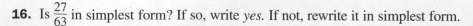

Spiral Review

16. Is $\frac{27}{63}$ in simplest form? If so, write *yes*. If not, rewrite it in simplest form.

17. What is the prime factorization of 48?

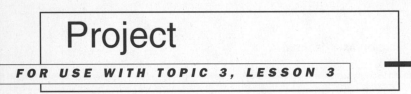

Warm-Ups

FOR USE WITH TOPIC 3, LESSON 3

Standardized Testing Warm-Ups

1. Which fraction is *not* equivalent to $\frac{4}{5}$?

   A $\frac{16}{20}$ **B** $\frac{80}{100}$ **C** $\frac{9}{10}$ **D** $\frac{800}{1000}$

2. Which decimal represents the number 586 ten-thousandths?

   A 0.586   **B** 0.0586   **C** 0.00586   **D**   0.000586

Homework Review Warm-Ups

Write each decimal as a fraction. Then write the fraction in simplest form if it is not already in simplest form.

3. 0.36   **4.** 0.4   **5.** 0.87   **6.** 0.22   **7.** 0.85

Project

FOR USE WITH TOPIC 3, LESSON 3

1. Suppose the school store is open for 50 hours this month. Whenever the store is open, a teacher supervisor is present. A partial schedule of the time teachers will help with the store this month is shown below. Complete the table.

Teacher	Hours present this month	Fractional part of time open this month	Decimal equivalent
Mr. Kim	3		
Ms. Hope	4		
Ms. Guerra	5		
Mr. Tonwe	8		

2. Teachers still need to be scheduled for the rest of the time the store is open this month. What fractional part of the time this month must still be scheduled? What is the decimal equivalent of this part of the time?

BRIDGE UNIT

Activity 3

FOR USE WITH TOPIC 3, LESSON 3

EXPLORATION

Cooperative Learning

Decimal Puzzles

In this activity, you will make puzzles showing fraction and decimal equivalents.
Then you will challenge your partner to put your puzzles together again.

You will need:

- plain paper
- scissors

What to do:

Work with a partner.

1. Have each partner cut a piece of paper into a simple 4-piece puzzle like the one shown to the right. Choose a fraction with a denominator of 4, 5, 20, 25, or 50. Write one of the following on each piece of the puzzle.

   - the fraction with the original denominator
   - an equivalent fraction with a denominator of 100
   - the decimal equivalent of the fraction
   - a sketch showing the fractional part of a square grid; you can trace an outline of a grid from a 100 square grid template.

2. Have each partner perform the steps above for 4 different fractions.

3. Have each partner mix all of her or his puzzle pieces. (Don't turn any pieces over.) Exchange puzzle pieces with your partner. Reassemble your partner's puzzles by finding equivalent values.

4. What strategies did you use to find fraction and decimal equivalents? Share your ideas with your partner.

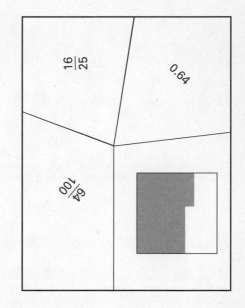

Basic Practice 3

Write the decimal equivalent for each fraction.

1. $\frac{4}{10}$
2. $\frac{1}{10}$
3. $\frac{23}{100}$
4. $\frac{4}{100}$

5. $\frac{1}{100}$
6. $\frac{100}{100}$
7. $\frac{82}{100}$
8. $\frac{918}{1000}$

9. $\frac{562}{1000}$
10. $\frac{615}{1000}$
11. $\frac{45}{1000}$
12. $\frac{2}{1000}$

Rewrite each fraction with a denominator of 10, 100, or 1000. Then write the decimal equivalent.

13. $\frac{1}{4}$
14. $\frac{1}{2}$
15. $\frac{3}{4}$
16. $\frac{2}{5}$

17. $\frac{6}{25}$
18. $\frac{3}{50}$
19. $\frac{5}{8}$
20. $\frac{1}{20}$

21. $\frac{2}{25}$
22. $\frac{17}{50}$
23. $\frac{9}{200}$
24. $\frac{23}{125}$

25. $\frac{11}{500}$
26. $\frac{15}{200}$
27. $\frac{3}{250}$
28. $\frac{124}{125}$

In Exercises 29–36, match each fraction with its decimal equivalent.

29. $\frac{7}{8}$ **A** 0.28

30. $\frac{7}{20}$ **B** 0.14

31. $\frac{7}{25}$ **C** 0.35

32. $\frac{7}{50}$ **D** 0.875

33. $\frac{3}{8}$ **E** 0.38

34. $\frac{9}{20}$ **F** 0.375

35. $\frac{9}{25}$ **G** 0.36

36. $\frac{19}{50}$ **H** 0.45

37. Write $\frac{4}{5}, \frac{4}{10}, \frac{4}{50}, \frac{4}{100}, \frac{4}{500}$, and $\frac{4}{1000}$ as decimals. Describe any patterns you see.

38. Write the decimal equivalent for the fraction $\frac{701}{10,000}$.

39. Rewrite the fraction $\frac{5}{16}$ with a denominator of 10,000. Use the fact that

10,000 ÷ 16 = 625. Then write the decimal equivalent for the fraction.

Working with Values Greater than 1

Goal
Find equivalent numbers using improper fractions, mixed numbers, and decimals.

You have been finding equivalent forms for decimals and fractions that are less than 1. You can also find equivalent decimals and fractions for numbers greater than 1.

Terms to Know	*Example / Illustration*
Improper fraction a fraction that is greater than or equal to 1; The numerator of an improper fraction is greater than or equal to the denominator.	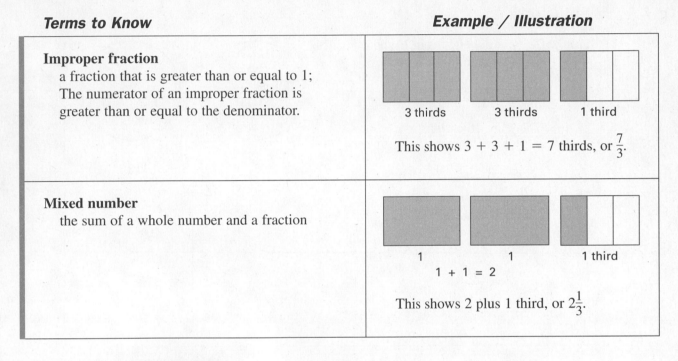 3 thirds 3 thirds 1 third This shows $3 + 3 + 1 = 7$ thirds, or $\frac{7}{3}$.
Mixed number the sum of a whole number and a fraction	1 1 1 third $1 + 1 = 2$ This shows 2 plus 1 third, or $2\frac{1}{3}$.

UNDERSTANDING THE MAIN IDEAS

You can see above that the mixed number $2\frac{1}{3}$ and the improper fraction $\frac{7}{3}$ show the same quantity. You can rewrite a mixed number as an equivalent improper fraction. To do this, first write the whole number as an improper fraction.

(continued)

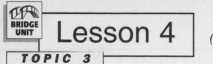
Example 1

Mr. Katz has 2 whole wheels and $\frac{5}{8}$ of a wheel of swiss cheese in the deli display case. He divides the whole amount into eighths of a wheel for packaging. Rewrite $2\frac{5}{8}$ as an improper fraction. How many packages of cheese will Mr. Katz have if each piece is $\frac{1}{8}$ of a wheel?

■ **Solution** ■

Step 1: There are 8 eighths in 1 whole. Multiply to find the number of eighths in 2 wholes.

$2 \times 8 = 16$

There are 16 eighths in 2 wholes.

$\frac{8}{8}$ $\frac{8}{8}$

Step 2: Add 5 eighths for the third cheese wheel.

$\frac{16}{8} + \frac{5}{8} = \frac{21}{8}$

So, $2\frac{5}{8} = \frac{21}{8}$.

$\frac{5}{8}$

Mr. Katz will have 21 packages of swiss cheese.

In Exercises 1–5, write an improper fraction equivalent to each mixed number.

1. $2\frac{2}{3}$ **2.** $1\frac{4}{5}$ **3.** $3\frac{1}{8}$ **4.** $5\frac{3}{4}$ **5.** $8\frac{1}{2}$

6. Trudy has $3\frac{1}{4}$ muffins. How many pieces can Trudy make if she cuts the muffins into fourths?

(continued)

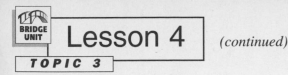

Lesson 4 *(continued)*

TOPIC 3

It is often easier to picture an improper fraction as a mixed number. To change an improper fraction to a mixed number, think about how many wholes the improper fraction contains.

Example 2

The PTA is sending out a newsletter. Each page of the newsletter will require one half of a ream of paper. This month's newsletter is 7 pages long. So, the PTA needs $\frac{7}{2}$ reams of paper. Written as a mixed number, how many reams will they use?

■ Solution ■

It takes 2 halves to make each whole. Divide to find the number of wholes in seven halves.

$7 \div 2 = 3$ R1 ← There are 3 whole reams, with 1 half ream left over.

The PTA will use $3\frac{1}{2}$ reams of paper for its newsletter.

In Exercises 7–11, write a mixed number equivalent to each improper fraction.

7. $\frac{9}{2}$ **8.** $\frac{22}{5}$ **9.** $\frac{15}{7}$ **10.** $\frac{17}{4}$ **11.** $\frac{11}{3}$

12. There were $\frac{17}{8}$ pizzas left after the party. Mr. Davies decided to keep whole pizzas only. How many whole pizzas did Mr. Davies save? What part did he throw away?

(continued)

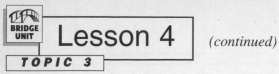

Improper fractions and mixed numbers also have decimal number equivalents.

Example 3

a. Write $\frac{13}{5}$ as a decimal.　　　　**b.** Write 5.25 as an improper fraction.

■ Solution ■

Step 1: a. Write $\frac{13}{5}$ as a mixed number.

$13 \div 5 = 2 \text{ R}3$

So, $\frac{13}{5} = 2\frac{3}{5}$.

b. Write 5.25 as a mixed number.

$0.25 = \frac{25}{100} = \frac{1}{4}$

So, $5.25 = 5\frac{1}{4}$.

Step 2: a. Write $2\frac{3}{5}$ as a decimal.

$\frac{3}{5} = \frac{6}{10} = 0.6$

So, $2\frac{3}{5} = 2.6$.

Then $\frac{13}{5} = 2\frac{3}{5} = 2.6$.

b. Write $5\frac{1}{4}$ as an improper fraction.

$5 \times 4 = 20 \text{ fourths}$

So, $5\frac{1}{4} = 21 \text{ fourths, or } \frac{21}{4}$.

Then $5.25 = 5\frac{1}{4} = \frac{21}{4}$.

Complete the table below.

	Decimal	Mixed number	Improper fraction
13.	1.1		
14.			$\frac{16}{5}$
15.		$2\frac{7}{8}$	
16.			$\frac{11}{4}$
17.	6.45		

··················
Spiral Review

18. Write $\frac{60}{170}$ in simplest form.

19. A train travels from Washington, D.C. to Philadelphia in 111 minutes. Does it take more or less than 2 hours? Explain.

Warm-Ups

FOR USE WITH TOPIC 3, LESSON 4

Standardized Testing Warm-Ups

1. Which improper fraction is equivalent to the mixed number $7\frac{5}{8}$?

A $\frac{75}{8}$ **B** $\frac{61}{8}$ **C** $\frac{56}{5}$ **D** $\frac{56}{8}$

2. Which one of the following numbers is *not* equivalent to the other three?

A $\frac{56}{100}$ **B** 5.6   **C** $5\frac{3}{5}$ **D** $\frac{28}{5}$

Homework Review Warm-Ups

Rewrite each fraction with a denominator of 10, 100, or 1000. Then write the equivalent decimal.

3. $\frac{5}{8}$ **4.** $\frac{2}{5}$ **5.** $\frac{3}{25}$ **6.** $\frac{13}{50}$ **7.** $\frac{17}{20}$

Project

FOR USE WITH TOPIC 3, LESSON 4

1. You can display only a part of your merchandise in the school store at one time. Suppose that for the items below, the first number represents your total supply. The second number represents the number of that item that will fit in its display space.

large binders: 41, 8   marker packages: 112, 15
mascot T-shirts: 70, 12   rulers: 95, 20

For each item, write an improper fraction for how many times the supply will fill an empty display. Rewrite the number as a mixed number in simplest form. Then tell how many times you can *completely refill* each display once it has sold out the first time.

2. To learn how to serve your customers more quickly, you record some of the wait times for checking out at the store. To analyze these times, it will be easier if they are in decimal number form. Write each time below as a decimal number.

a. 3 min 21 s   **b.** 1 min 15 s   **c.** 4 min 48 s   **d.** 3 min 51 s

Activity 4

PROBLEM SOLVING/APPLICATION

Cooperative Learning

Relating Travel Times and Distances

Did you know that the cruising speed of some passenger jets is about 600 miles per hour? In this activity, you and a partner will use flight times to estimate the distance between cities.

What you need:

• a calculator (optional)

What to do:

Work with a partner.

1. The table below gives some flight times. Rewrite each time as indicated, and have your partner estimate the distance. Reverse roles and repeat the process to check your answers. To help you get started, an example is given below.

   Suppose the flight time from Las Vegas to Atlanta is 3 h 24 min.

   mixed number: $3 \text{ h } 24 \text{ min} = 3\frac{24}{60} \text{ h} = 3\frac{2}{5} \text{ h}$

   decimal number: $3\frac{2}{5} \text{ h} = 3.4 \text{ h}$

   estimated distance: $3.4 \times 600 = 3.4 \times 100 \times 6 = 340 \times 6 = 2040 \text{ mi}$

Cities of travel	Time	Time (whole or mixed number)	Time (decimal)	Estimated distance (mi)
Los Angeles to Pittsburgh	4 h			
Los Angeles to Boston	5 h			
Tulsa to Cleveland	1 h 30 min			
New Orleans to Cleveland	1 h 45 min			
Denver to Dallas	1 h 15 min			
Denver to Atlanta	2 h 27 min			

2. The distance between Houston and Philadelphia is about 1500 miles. Estimate the duration of the flight as a mixed number and as hours and minutes. To find time, you can divide distance by speed. Discuss how you found your answers.

Basic Practice 4

FOR USE WITH TOPIC 3, LESSON 4

Write an improper fraction equivalent to each mixed number.

1. $3\frac{1}{3}$ **2.** $3\frac{1}{2}$ **3.** $3\frac{3}{4}$ **4.** $7\frac{5}{6}$

5. $3\frac{2}{5}$ **6.** $6\frac{3}{10}$ **7.** $2\frac{2}{5}$ **8.** $4\frac{5}{7}$

Write a mixed number equivalent to each improper fraction. Write your answer in simplest form.

9. $\frac{25}{8}$ **10.** $\frac{28}{6}$ **11.** $\frac{17}{3}$ **12.** $\frac{24}{5}$

13. $\frac{15}{4}$ **14.** $\frac{13}{10}$ **15.** $\frac{4}{3}$ **16.** $\frac{46}{10}$

Write a decimal equivalent to each mixed number or improper fraction.

17. $1\frac{19}{20}$ **18.** $\frac{67}{10}$ **19.** $8\frac{1}{2}$ **20.** $\frac{101}{100}$

21. $\frac{81}{10}$ **22.** $\frac{21}{6}$ **23.** $\frac{75}{4}$ **24.** $11\frac{1}{4}$

In Exercises 25–32, write an improper fraction equivalent to each decimal. Write your answer in simplest form.

25. 12.4   **26.** 3.25   **27.** 2.1   **28.** 6.15

29. 9.5   **30.** 5.24   **31.** 1.35   **32.** 2.625

33. Each number below corresponds to a point on the number line shown. Give the letters in the order that they match the points from left to right.

A $1\frac{2}{5}$ **B** $2\frac{1}{2}$ **C** 2.75   **D** $\frac{5}{4}$ **E** $\frac{10}{3}$ **F** 3.05

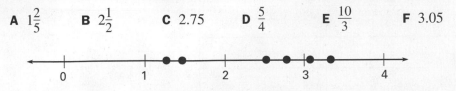

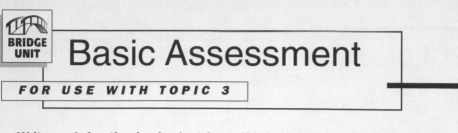

Write each fraction in simplest form. If it is already in simplest form, write *simplest form.*

1. $\frac{25}{80}$ **2.** $\frac{23}{46}$ **3.** $\frac{15}{27}$ **4.** $\frac{34}{51}$

5. $\frac{9}{16}$ **6.** $\frac{10}{14}$ **7.** $\frac{87}{90}$ **8.** $\frac{13}{30}$

Write an equivalent fraction for each decimal. Then write the fraction in simplest form if it is not already in simplest form.

9. 0.12   **10.** 0.61   **11.** 0.84   **12.** 0.34

Write each fraction as a decimal.

13. $\frac{3}{4}$ **14.** $\frac{4}{5}$ **15.** $\frac{1}{2}$ **16.** $\frac{5}{8}$

17. $\frac{9}{20}$ **18.** $\frac{7}{10}$ **19.** $\frac{22}{25}$ **20.** $\frac{19}{50}$

Write a decimal and an improper fraction for each mixed number.

	Improper fraction	Mixed number	Decimal
21.		$2\frac{7}{10}$	
22.		$3\frac{3}{8}$	
23.		$4\frac{3}{5}$	
24.		$5\frac{17}{20}$	
25.		$3\frac{2}{25}$	
26.		$1\frac{24}{25}$	
27.		$8\frac{5}{20}$	
28.		$2\frac{13}{100}$	

29. A train travels from Oakmont to Elmer in 225 minutes. How long does the trip take? Write your answer in terms of hours and minutes and as a mixed number in simplest form.

30. Compare changing a mixed number to an improper fraction and changing an improper fraction to a mixed number. What pattern do you see? How can you explain it?

Lesson 1

Lesson Exercises *pages 76–77*

1. No; $\frac{1}{2}$　　**2.** No; $\frac{3}{4}$　　**3.** No; $\frac{3}{5}$

4. Yes.　　**5.** No; $\frac{2}{3}$　　**6.** No; $\frac{2}{5}$

7. Yes.　　**8.** No; $\frac{1}{2}$　　**9.** $\frac{3}{7}$

10. $\frac{1}{3}$　　**11.** $\frac{3}{5}$　　**12.** $\frac{1}{3}$

13. $\frac{2}{3}$　　**14.** $\frac{3}{8}$　　**15.** $\frac{7}{8}$

16. simplest form　　**17.** 2 and 3

18. prime factorization

Warm-Ups *page 78*

1. C　　**2.** B　　**3.** 41, 43, 47

4. $2 \times 3 \times 3 \times 5$

Project *page 78*

1. $\frac{15}{60}, \frac{20}{60}, \frac{25}{60}, \frac{30}{60}, \frac{35}{60}, \frac{40}{60}, \frac{45}{60}$

2. $\frac{1}{4}, \frac{1}{3}, \frac{5}{12}, \frac{1}{2}, \frac{7}{12}, \frac{2}{3}, \frac{3}{4}$

3. Answers will vary, but Lisa and Benjamin must be scheduled after LaToya, and Renee and Malcolm must be scheduled before LaToya.

Activity *page 79*

1–2. Check students' work.

3. a. $\frac{1}{2}$　　**b.** $\frac{1}{3}$　　**c.** $\frac{1}{4}$

d. $\frac{2}{3}$　　**e.** $\frac{1}{2}$　　**f.** $\frac{3}{4}$

4. The pieces cover the same area. For example, 2 one-fourth circle pieces cover the same area as 1 one-half circle piece.

5. Possible answer: As few pieces as possible were used to model the fraction.

Basic Practice *page 80*

1. No.　　**2.** Yes.　　**3.** Yes.

4. No.　　**5.** No.　　**6.** Yes.

7. No.　　**8.** Yes.　　**9.** No.

10. No.　　**11.** Yes.　　**12.** No.

13. simplest form　　**14.** simplest form

15. $\frac{3}{10}$　　**16.** $\frac{1}{4}$　　**17.** $\frac{1}{4}$

18. $\frac{4}{7}$　　**19.** $\frac{1}{6}$　　**20.** simplest form

21. $\frac{17}{25}$　　**22.** $\frac{3}{4}$　　**23.** $\frac{3}{5}$

24. $\frac{2}{5}$　　**25.** simplest form

26. $\frac{3}{8}$　　**27.** simplest form

28. $\frac{3}{8}$　　**29.** $\frac{8}{15}$　　**30.** $\frac{2}{3}$

31. $\frac{2}{3}$　　**32.** $\frac{2}{5}$　　**33.** $\frac{4}{7}$

34. $\frac{1}{3}$　　**35.** simplest form

36. simplest form　　**37.** $\frac{4}{9}$

38. $\frac{1}{3}$　　**39.** $\frac{2}{9}$　　**40.** $\frac{2}{3}$

Lesson 2

Lesson Exercises *page 82*

1. $\frac{4}{10}, \frac{2}{5}$　　**2.** $\frac{1}{10}$　　**3.** $\frac{9}{10}$

4. $\frac{7}{10}$　　**5.** $\frac{3}{10}$　　**6.** $\frac{5}{10}, \frac{1}{2}$

7. $\frac{8}{10}, \frac{4}{5}$　　**8.** $\frac{2}{10}, \frac{1}{5}$　　**9.** $\frac{40}{100}, \frac{2}{5}$

10. $\frac{15}{100}, \frac{3}{20}$　　**11.** $\frac{28}{100}, \frac{7}{25}$　　**12.** $\frac{75}{100}, \frac{3}{4}$

13. $\frac{9}{100}$　　**14.** $\frac{18}{100}, \frac{9}{50}$　　**15.** $\frac{65}{100}, \frac{13}{20}$

16. $\frac{5}{100}; \frac{1}{20}$ **17.** $\frac{12}{100}; \frac{3}{25}$ **18.** $\frac{68}{100}; \frac{17}{25}$

19. $\frac{37}{100}$ **20.** $\frac{1}{100}$

21. Sara; 169 cm = 1.69 m, which is greater than 1.65 m.

22. $\frac{3}{4}$ of an hour

Warm-Ups *page 83*

1. C   **2.** B   **3.** B

4. E   **5.** C   **6.** D

7. A

Project *page 83*

1. $\frac{3}{4}, \frac{14}{25}, \frac{2}{5}, \frac{1}{4}, \frac{1}{25}$ **2.** 2

3. $\frac{3}{4}$ and $\frac{1}{4}$ (or 0.75 and 0.25); Sample answer: If a student answers "somewhat satisfied" or better to at least 4 questions, then that student must have answered "unsatisfied" or worse to *no more than* 1 question. So, the two possibilities represented by $\frac{3}{4}$ and $\frac{1}{4}$ make up the whole portion of answers without overlapping. The whole portion is represented by 1.

Activity *page 84*

1. Check students' work.

2. a. $\frac{1}{10}$ **b.** $\frac{1}{4}$ **c.** $\frac{8}{25}$

   d. $\frac{1}{2}$ **e.** $\frac{3}{5}$ **f.** $\frac{3}{4}$

   g. $\frac{4}{5}$ **h.** $\frac{9}{10}$

Possible answer: $\frac{1}{10}$ and $\frac{9}{10}$; they are close to 0 and to 1.

3. Answers will vary.

4. a. Students may say they used visual clues, estimated, or counted.

   b. Possible answer: $\frac{1}{4}, \frac{1}{2},$ and $\frac{3}{4}$; These are easier to see because these fractions are used very frequently in everyday life, such as in cooking.

   c. Possible answer: $\frac{8}{25}$ and $\frac{3}{5}$; These may be harder to recognize because you don't see examples of these as often in everyday life.

Basic Practice *page 85*

1. $\frac{6}{10}; \frac{3}{5}$ **2.** $\frac{60}{100}; \frac{3}{5}$ **3.** $\frac{2}{100}; \frac{1}{50}$

4. $\frac{4}{100}; \frac{1}{25}$ **5.** $\frac{96}{100}; \frac{24}{25}$ **6.** $\frac{55}{100}; \frac{11}{20}$

7. $\frac{84}{100}; \frac{21}{25}$ **8.** $\frac{33}{100}$ **9.** 0.38; $\frac{38}{100}; \frac{19}{50}$

10. 0.62; $\frac{62}{100}; \frac{31}{50}$ **11.** 0.24; $\frac{24}{100}; \frac{6}{25}$

12. 0.36; $\frac{36}{100}; \frac{9}{25}$

13. 0.25   **14.** 0.05

15. a. $\frac{3}{20}$ **b.** 3

16. $\frac{125}{1000}$ and $\frac{375}{1000}; \frac{1}{8}$ and $\frac{3}{8}$

Lesson 3

Lesson Exercises *pages 86–87*

1. 0.3   **2.** 0.9   **3.** 0.27

4. 0.81   **5.** 0.123   **6.** $\frac{4}{10};$ 0.4

7. $\frac{5}{10};$ 0.5   **8.** $\frac{94}{100};$ 0.94

9. $\frac{8}{100};$ 0.08   **10.** $\frac{16}{100};$ 0.16

11. $\frac{45}{100};$ 0.45   **12.** $\frac{52}{100};$ 0.52

13. $\frac{615}{1000};$ 0.615   **14.** $\frac{556}{1000};$ 0.556

15. $\frac{625}{1000}$; 0.625   **16.** $\frac{3}{7}$

17. $2 \times 2 \times 2 \times 2 \times 3$

Warm-Ups *page 88*

1. C   **2.** C   **3.** $\frac{36}{100}$; $\frac{9}{25}$

4. $\frac{4}{10}$; $\frac{2}{5}$ **5.** $\frac{87}{100}$ **6.** $\frac{22}{100}$; $\frac{11}{50}$

7. $\frac{85}{100}$; $\frac{17}{20}$

Project *page 88*

1. Mr Kim: $\frac{3}{50} = \frac{6}{100}$, 0.06;

Ms. Hope: $\frac{4}{50} = \frac{8}{100}$, 0.8;

Ms. Guerra: $\frac{5}{50} = \frac{1}{10}$, 0.1;

Mr. Tonwe: $\frac{8}{50} = \frac{16}{100}$, 0.16

2. $\frac{30}{50} = \frac{3}{5} = \frac{6}{10}$; 0.6

Activity 3 *page 89*

1–4. Answers will vary. Check students' work.

Basic Practice *page 90*

1. 0.4   **2.** 0.1   **3.** 0.23

4. 0.04   **5.** 0.01   **6.** 1.00

7. 0.82   **8.** 0.918   **9.** 0.562

10. 0.615   **11.** 0.045   **12.** 0.002

13. $\frac{25}{100}$; 0.25   **14.** $\frac{5}{10}$; 0.5   **15.** $\frac{75}{100}$; 0.75

16. $\frac{4}{10}$; 0.4   **17.** $\frac{24}{100}$; 0.24   **18.** $\frac{6}{100}$; 0.06

19. $\frac{625}{1000}$; 0.625   **20.** $\frac{5}{100}$; 0.05

21. $\frac{8}{100}$; 0.08   **22.** $\frac{34}{100}$; 0.34

23. $\frac{45}{1000}$; 0.045   **24.** $\frac{184}{1000}$; 0.184

25. $\frac{22}{1000}$; 0.022   **26.** $\frac{75}{1000}$; 0.075

27. $\frac{12}{1000}$; 0.012   **28.** $\frac{992}{1000}$; 0.992

29. D   **30.** C   **31.** A

32. B   **33.** F   **34.** H

35. G   **36.** E

37. 0.8, 0.4, 0.08, 0.04, 0.008, 0.004; Possible pattern: As the denominators increase, the decimal equivalents decrease.

38. 0.0701   **39.** $\frac{3125}{10,000}$; 0.3125

Lesson 4

Lesson Exercises *pages 92–94*

1. $\frac{8}{3}$ **2.** $\frac{9}{5}$ **3.** $\frac{25}{8}$

4. $\frac{23}{4}$ **5.** $\frac{17}{2}$ **6.** 13 pieces

7. $4\frac{1}{2}$ **8.** $4\frac{2}{5}$ **9.** $2\frac{1}{7}$

10. $4\frac{1}{4}$ **11.** $3\frac{2}{3}$

12. 2 pizzas; $\frac{1}{8}$ pizza

	Decimal	Mixed number	Improper fraction
13.	1.1	$1\frac{1}{10}$	$\frac{11}{10}$
14.	3.2	$3\frac{1}{5}$	$\frac{16}{5}$
15.	2.875	$2\frac{7}{8}$	$\frac{23}{8}$
16.	2.75	$2\frac{3}{4}$	$\frac{11}{4}$
17.	6.45	$6\frac{9}{20}$	$\frac{129}{20}$

18. $\frac{6}{17}$

19. less than 2 hours; 2 hours = 120 minutes, and 111 < 120.

Warm-Ups *page 95*

1. B   **2.** A   **3.** $\frac{625}{1000}$; 0.625

4. $\frac{4}{10}$; 0.4   **5.** $\frac{12}{100}$; 0.12   **6.** $\frac{26}{100}$; 0.26

7. $\frac{85}{100}$; 0.85

Project *page 95*

1. binders: $\frac{41}{8}$; $5\frac{1}{8}$; 4

markers: $\frac{112}{15}$; $7\frac{7}{15}$; 6

T-shirts: $\frac{70}{12}$; $5\frac{5}{6}$; 4

rulers: $\frac{95}{20}$; $4\frac{3}{4}$; 3

2. a. 3.35 min   **b.** 1.25 min

       c. 4.8 min   **d.** 3.85 min

Activity 4 *page 96*

Cities of travel	Time	Time (whole or mixed number)	Time (decimal)	Estimated distance (mi)
Los Angeles to Pittsburgh	4 h	4 h	4.0 h	2400 mi
Los Angeles to Boston	5 h	5 h	5.0 h	3000 mi
Tulsa to Cleveland	1 h 30 min	$1\frac{1}{2}$ h	1.5 h	900 mi
New Orleans to Cleveland	1 h 45 min	$1\frac{3}{4}$ h	1.75 h	1050 mi
Denver to Dallas	1 h 15 min	$1\frac{1}{4}$ h	1.25 h	750 mi
Denver to Atlanta	2 h 27 min	$2\frac{9}{20}$ h	2.45 h	1470 mi

2. about $2\frac{1}{2}$ hours, or 2 h 30 min; Strategies will vary.

Basic Practice *page 97*

1. $\frac{10}{3}$ **2.** $\frac{7}{2}$ **3.** $\frac{15}{4}$

4. $\frac{47}{6}$ **5.** $\frac{17}{5}$ **6.** $\frac{63}{10}$

7. $\frac{12}{5}$ **8.** $\frac{33}{7}$ **9.** $3\frac{1}{8}$

10. $4\frac{2}{3}$ **11.** $5\frac{2}{3}$ **12.** $4\frac{4}{5}$

13. $3\frac{3}{4}$ **14.** $1\frac{3}{10}$ **15.** $1\frac{1}{3}$

16. $4\frac{3}{5}$ **17.** 1.95   **18.** 6.7

19. 8.5   **20.** 1.01   **21.** 8.1

22. 3.5   **23.** 18.75   **24.** 11.25

25. $\frac{62}{5}$ **26.** $\frac{13}{4}$ **27.** $\frac{21}{10}$

28. $\frac{123}{20}$ **29.** $\frac{19}{2}$ **30.** $\frac{131}{25}$

31. $\frac{27}{20}$ **32.** $\frac{21}{8}$

33. D, A, B, C, F, E

Topic 3 Basic Assessment *page 98*

1. $\frac{5}{16}$ **2.** $\frac{1}{2}$ **3.** $\frac{5}{9}$

4. $\frac{2}{3}$ **5.** simplest form

6. $\frac{5}{7}$ **7.** $\frac{29}{30}$ **8.** simplest form

9. $\frac{12}{100}; \frac{3}{25}$ **10.** $\frac{61}{100}$ **11.** $\frac{84}{100}; \frac{21}{25}$

12. $\frac{34}{100}; \frac{17}{50}$ **13.** 0.75   **14.** 0.8

15. 0.5   **16.** 0.625   **17.** 0.45

18. 0.7   **19.** 0.88   **20.** 0.38

	Improper fraction	Mixed number	Decimal
21.	$\frac{27}{10}$	$2\frac{7}{10}$	2.7
22.	$\frac{27}{8}$	$3\frac{3}{8}$	3.375
23.	$\frac{23}{5}$	$4\frac{3}{5}$	4.6
24.	$\frac{117}{20}$	$5\frac{17}{20}$	5.85
25.	$\frac{77}{25}$	$3\frac{2}{25}$	3.08
26.	$\frac{49}{25}$	$1\frac{24}{25}$	1.96
27.	$\frac{165}{20}$	$8\frac{5}{20}$	8.25
28.	$\frac{213}{100}$	$2\frac{13}{100}$	2.13

29. 3 h 45 min, or $3\frac{3}{4}$ hours

30. Possible answer: You multiply to change a mixed number to an improper fraction, and you divide to change an improper fraction to a mixed number. Multiplication and division "undo" each other. You can think of changing between mixed numbers and improper fractions as each undoing the other.

Identifying Quadrilaterals

Goal
Identify the special
features of
quadrilaterals.
Identify and sketch
quadrilaterals by
description.

Tennis courts, soccer fields, and baseball diamonds are just a few of the four-sided sports playing areas. Quadrilaterals may all have four sides, but they are not all alike. In this lesson, you'll learn about some of the special quadrilaterals and their characteristics.

Terms to Know	*Example / Illustration*
Polygon a closed figure, in a plane, formed by line segments joined only at their endpoints	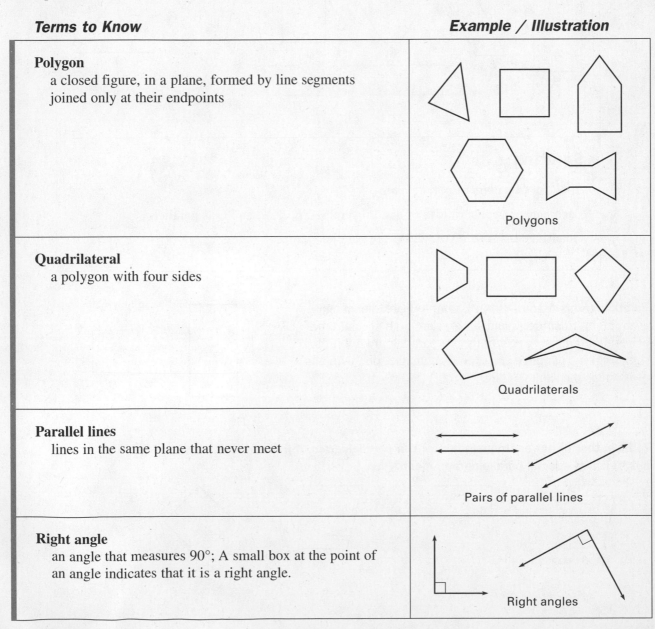 Polygons
Quadrilateral a polygon with four sides	Quadrilaterals
Parallel lines lines in the same plane that never meet	Pairs of parallel lines
Right angle an angle that measures 90°; A small box at the point of an angle indicates that it is a right angle.	Right angles

(continued)

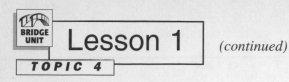

UNDERSTANDING THE MAIN IDEAS

One way to identify quadrilaterals is to look for parallel sides. A **parallelogram** is a quadrilateral that has two pairs of parallel sides.

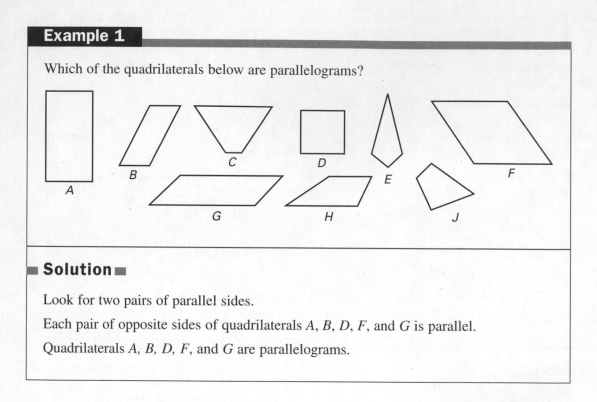

Example 1

Which of the quadrilaterals below are parallelograms?

■ **Solution** ■

Look for two pairs of parallel sides.

Each pair of opposite sides of quadrilaterals *A*, *B*, *D*, *F*, and *G* is parallel.

Quadrilaterals *A*, *B*, *D*, *F*, and *G* are parallelograms.

Parallelograms *A* and *D* from Example 1 are shown again at the right. *Rectangles* and *squares* are both special types of parallelograms. A **rectangle** is a parallelogram with four right angles. A **square** is a special rectangle, one with all of its sides the same length.

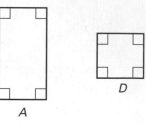

Tell whether or not each quadrilateral is a parallelogram. If it is not, tell why not. If it is a special parallelogram, identify it.

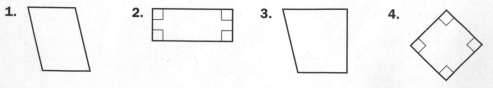

1.   2.   3.   4.

(continued)

Another type of special parallelogram is called a *rhombus*. A rhombus is a parallelogram whose four sides are all the same length.

Example 2

Which of the quadrilaterals below are rhombuses?

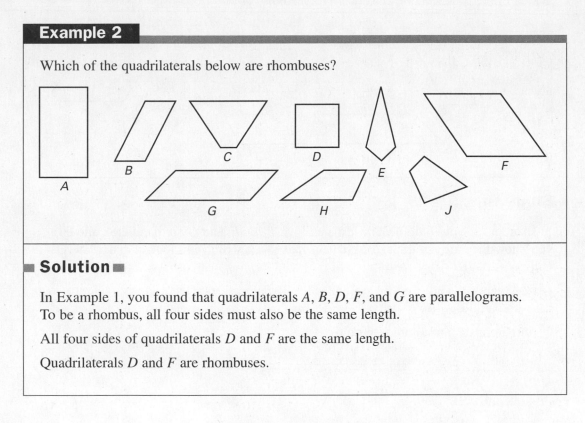

■ Solution ■

In Example 1, you found that quadrilaterals *A*, *B*, *D*, *F*, and *G* are parallelograms. To be a rhombus, all four sides must also be the same length.

All four sides of quadrilaterals *D* and *F* are the same length.

Quadrilaterals *D* and *F* are rhombuses.

You saw in Example 1 that a square is a special kind of parallelogram. In Example 2, you see that a square is also a special type of rhombus. A square is a rhombus with four right angles.

Tell whether or not each quadrilateral is a rhombus. If it is not, tell why not. If it is a special rhombus, identify it.

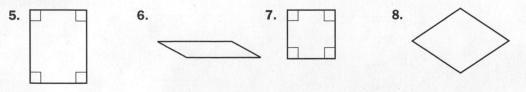

A *trapezoid* is not a parallelogram, but it is a special type of quadrilateral. A **trapezoid** is a quadrilateral that has *exactly one* pair of parallel sides.

(continued)

Example 3

Which of the quadrilaterals below are trapezoids?

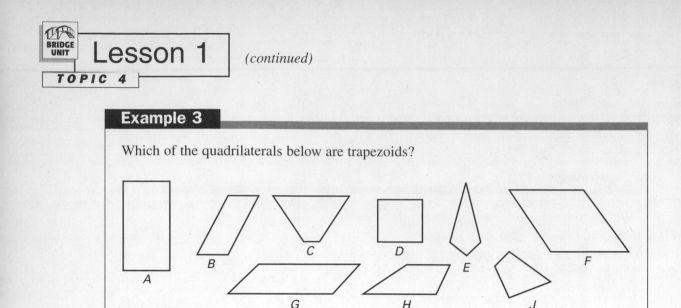

Solution

In Example 1, you found that quadrilaterals A, B, D, F, and G are parallelograms, so you know they are not trapezoids. This means you only have to look at quadrilaterals C, E, H, and J.

Quadrilaterals C and H each have one pair of parallel sides. Quadrilaterals E and J have no parallel sides.

Quadrilaterals C and H are trapezoids.

Answer each question.

9. How are a square and a rectangle different?

10. How are a parallelogram and a rhombus different?

11. How are a square and a rhombus alike?

12. How is a trapezoid different from the other special quadrilaterals?

13. Write as many different quadrilateral names as you can for each figure.

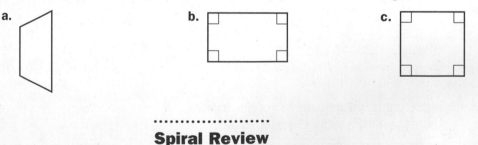

a.   b.   c.

Spiral Review

14. Jené loads 208 cases of frozen potatoes onto the refrigerator truck. Each case holds 48 packages of potatoes. About how many packages of potatoes does she load?

15. By which of the numbers 2, 3, 5, 9, and 10 is 379,980 divisible?

BRIDGE UNIT

Warm-Ups

FOR USE WITH TOPIC 4, LESSON 1

Standardized Testing Warm-Ups

1. Which of the following can *never* be a rhombus?

   A parallelogram   **B** square   **C** rectangle   **D** trapezoid

2. Which of the following is *not* a polygon?

   A trapezoid   **B** triangle   **C** circle   **D** rhombus

Homework Review Warm-Ups

Write each improper fraction as a mixed number and an equivalent decimal.

3. $\frac{5}{4}$ **4.** $\frac{11}{10}$ **5.** $\frac{27}{20}$ **6.** $\frac{76}{25}$

Project

FOR USE WITH TOPIC 4, LESSON 1

1. Sketch a floor plan for the school store. Use grid or dot paper. Use your room measurements from the project from Topic 1, Lesson 3 to give your drawing the proper dimensions. Use a simple scale, like letting one side of a grid square represent 1 meter. Remember to include the following.

- Provide enough space for displaying goods.
- Provide floor space for customers to walk.
- Provide space for store workers to assist customers and take payment for purchases.

2. Plan the product display areas for your store.

- If there are tables, shelves, or bookcases in your classroom, use them as models for the displays in your floor plan.
- If your classroom contains mostly desks, you might want to model displays made by pushing desks together.
- Include measurements for the lengths and widths of the floor spaces, tables, and so on.

3. Identify any types of quadrilaterals that you have used or can find in your plans. Have you shown any other polygons that are not quadrilaterals?

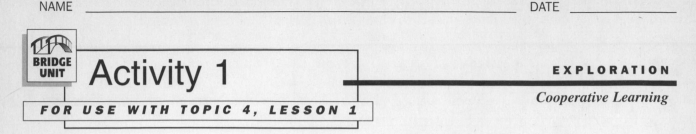

BRIDGE UNIT

Activity 1

FOR USE WITH TOPIC 4, LESSON 1

Quadrilateral Puzzles

A *tangram* is an ancient puzzle invented in China. The pieces of the puzzle are called *tans*. The finished arrangements made from the pieces are called *tangrams*. In this activity, you will use tans to form quadrilaterals.

You will need:

- a tangram template
- scissors
- plain paper

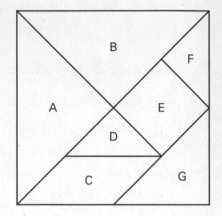

What to do:

Work with a partner.

1. Trace the tangram from the template onto a sheet of unlined paper. Label each piece as shown, writing each label on both sides of the paper. Cut along the lines so that you have seven pieces.

2. Use the labeled tans to form quadrilateral tangrams as described below. It is okay to flip a piece over if it helps.

   a. Use tans A, D, F, and G to form a square.

   b. Use tans A, D, F, and G to form a parallelogram that is not a square.

   c. Use tans C, D, E, and F to form a rectangle.

   d. Use tans D, E, and F to form a trapezoid.

   e. Use tans D and F to form a parallelogram that is not a squarc.

   f. Use tans D, F, and G to form a square.

3. Without showing your partner, form two more quadrilaterals using different groups of tans. Tell your partner which pieces you used and the type of quadrilateral you created. Then challenge your partner to recreate your figure.

4. Working together, try to make a parallelogram (not a square or rectangle) using all seven tans.

Basic Practice 1

BRIDGE UNIT

Classify each statement as *true* or *false*.

1. Every rectangle is a parallelogram.

2. Every parallelogram is a square.

3. Every square is a rhombus.

4. Every rhombus is a parallelogram.

5. Every trapezoid is a rectangle.

6. Some trapezoids are rectangles.

7. Some rhombuses are squares.

8. Some parallelograms are trapezoids.

9. Some rectangles are rhombuses.

Solve each riddle.

10. I am a quadrilateral with two pairs of parallel sides and four sides of the same length. All of my angles are the same measure, too. What am I?

11. I am a quadrilateral with two pairs of parallel sides. All of my angles are the same measure, but my sides are not all the same length. What am I?

12. I am a quadrilateral with exactly one pair of parallel sides. What am I?

13. I am any quadrilateral with two pairs of parallel sides. What am I?

Use rectangular dot paper to sketch each figure. Make your figure fit as few other special quadrilateral names as possible.

14. rectangle

15. parallelogram

16. trapezoid

17. rhombus

18. Evan said, "Every rectangle is a square." Joan said, "No, you're wrong. Every square is a rectangle." Who is right? Explain your answer.

19. What is the fewest number of figures you would have to draw to display a square, a rhombus, a rectangle, a parallelogram, and a trapezoid? What are the figures?

Goal
Use a protractor to measure angles and to draw angles of specified measures.

Measuring and Constructing Angles

Whenever two sides of a quadrilateral or other polygon meet, they meet in an angle. Sometimes the angle is a right angle, as for rectangles and squares. But whatever the angle is, you can measure it. Then you can start to look for patterns in the angles.

Terms to Know	*Example / Illustration*
Ray a part of a line that has one endpoint and extends in one direction without ending	A B This is ray AB, or $\overrightarrow{AB}$. Its endpoint is A.
Angle a figure formed by two rays that have a common endpoint; The rays form the *sides* of the angle. The endpoint is called the *vertex* of the angle. The symbol for angle is $\angle$.	C A B This is angle A, or $\angle A$. Rays $\overrightarrow{AB}$ and $\overrightarrow{AC}$ form its sides. Its vertex is the point A.
Protractor a tool that you use to measure angles; Protractors measure angles in units called *degrees* (°).	The measure of $\angle N$ is 70 degrees, or 70°.
Diagonal a line segment that joins two vertices of a polygon, but is not a side of the polygon; (Since a corner of a polygon forms an angle, each corner point is called a vertex—vertices is the plural).	A B C D Segment AD, or $\overline{AD}$, is a diagonal of rectangle $ABCD$.

(continued)

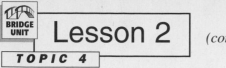

UNDERSTANDING THE MAIN IDEAS

Most protractors have two scales. The scale starting at the left edge is read clockwise from 0° to 180°. The scale starting at the right edge is read counterclockwise from 0° to 180°.

Example 1

Use a protractor to find the measure of ∠X.

■ Solution ■

Step 1: Place the protractor's center on the vertex of the angle.

Step 2: Line up the protractor's 0° line with one ray of the angle.

Step 3: Read the measure on the protractor where the other ray crosses it. Because you are using the 0° measure on the left, read the clockwise scale.

The measure of ∠X is 55°.

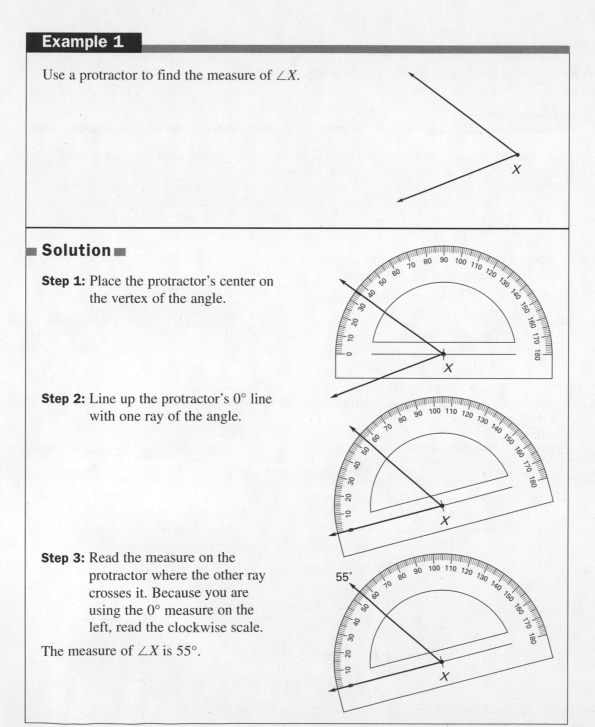

(continued)

Use a protractor to measure each angle to the nearest 5°.

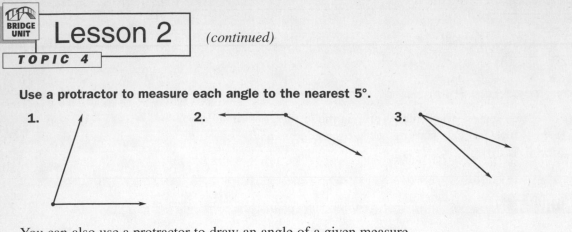

1.

2.

3.

You can also use a protractor to draw an angle of a given measure.

Example 2

A wheelchair ramp being installed at a curb meets the ground at an angle of about 5°. Draw a 5° angle to represent the ramp. Label it ∠B.

■ Solution ■

Step 1: Draw a ray. Because the angle is formed by the ramp meeting the ground, draw a horizontal ray to represent the ground.

Step 2: Place the center point of the protractor on the endpoint of the ray, with the 0° mark along the ray. Mark a point at the edge of the protractor at the measure you want.

Step 3: Use a straightedge to draw a ray connecting the endpoint of the first ray with the new point. Label the angle *B* at the vertex.

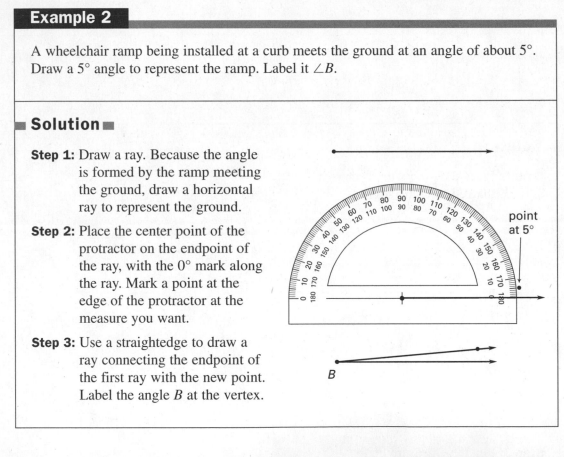

point at 5°

B

Use a protractor to draw an angle of the given measure.

4. 30°   **5.** 140°   **6.** 75°   **7.** 45°

· · · · · · · · · · · · · · · · · · · ·

Spiral Review

8. Use a factor tree to find the prime factorization of 128.

9. Give the compatible numbers you would use to estimate the quotient 2691 ÷ 29. Then write your estimate.

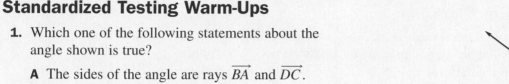

Warm-Ups

BRIDGE UNIT

FOR USE WITH TOPIC 4, LESSON 2

Standardized Testing Warm-Ups

1. Which one of the following statements about the angle shown is true?

   A The sides of the angle are rays $\overrightarrow{BA}$ and $\overrightarrow{DC}$.

   B The vertices of the angle are A and C.

   C The best one-letter name for the angle is $\angle D$.

   D Ray $\overrightarrow{DA}$ is the diagonal of the angle.

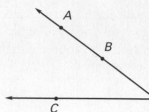

2. Which one of the following statements is *not* true?

   A The point where the sides of an angle meet is called the vertex.

   B The diagonal of a polygon is any segment that connects two vertices of the polygon.

   C A ray has only one endpoint.

   D The points where the sides of a polygon meet are called its vertices.

Homework Review Warm-Ups

3. How are a rhombus and a square alike? How are they different?

4. What is the name for a quadrilateral that has one and only one pair of parallel sides? Do any of the sides of this quadrilateral have to have the same length?

Project

FOR USE WITH TOPIC 4, LESSON 2

1. Review the decisions you have made so far regarding the school store. Think about the topics below. Now that you have spent more time on your plans, is there anything you want to change? Explain.

   a. hours of operation and work schedules

   b. items to be sold

   c. store floor plan and displays

   d. cost to buy each item

   e. selling cost of each item

   f. predicted profit from each item

2. Create an advertisement for the school store.

   a. What do you need to include in the advertisement?

   b. What needs to be calculated but not advertised?

BRIDGE UNIT

Activity 2

FOR USE WITH TOPIC 4, LESSON 2

Angle Measures in Polygons

Every quadrilateral has four corners, or vertices. If you draw a diagonal between two of the vertices, the diagonal will divide the quadrilateral into two triangles. Using a protractor to measure the angles in the triangles, you can find patterns in the angle measures.

You will need:

- a protractor
- a ruler

What to do:

1. Use a ruler to draw four different types of quadrilaterals. Identify each quadrilateral by its type. Make sure that if you draw a square or a rhombus, all four sides are of equal length. Use letters to name the vertices of each quadrilateral.

2. Draw a diagonal for each quadrilateral. Remember, a diagonal is a line segment that joins two vertices of a polygon, but is not a side of the polygon. Number the six angles that are formed from 1–6, as shown in quadrilateral *QRST* at the right.

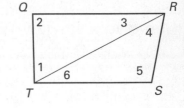

3. Use a protractor to measure each of the six angles to the nearest degree. Record your findings in a table like the one below. In the last column, record the sum of the measures of the six angles in each quadrilateral.

Quadrilateral	∠1	∠2	∠3	∠4	∠5	∠6	Sum

4. Answer the following questions.

   a. Find the sum of the measures of the three angles in each triangle. What do you find?

   b. Compare the sums of the angle measures in each triangle with the sums of the angle measures in each quadrilateral. What do you find? Why does this make sense?

   c. Suppose you did not draw a diagonal, and simply measured each of the four angles of the quadrilateral. What would you predict as the sum of the measures of the angles? Explain your thinking.

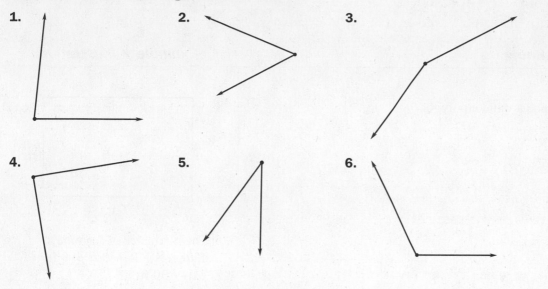

BRIDGE UNIT

Basic Practice 2

FOR USE WITH TOPIC 4, LESSON 2

Find the measure of each angle to the nearest 5°.

1.

2.

3.

4.

5.

6.

Use a protractor to draw an angle with the given measure.

7. 50°   **8.** 140°   **9.** 105°   **10.** 25°

Find the measure of each angle to the nearest 5°. Then find the sum of the angle measures.

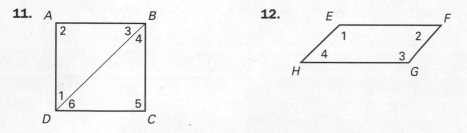

11.

A   B
2   3 / 4

1 / 6   5
D   C

12.

E   F
1   2
4   3
H   G

13. Draw a triangle that has two angles measuring 30° each. What is the measure of the third angle? Tell how you know.

14. Draw a quadrilateral that has three angles measuring 90° each. What is the measure of the fourth angle? Tell how you know.

Goal
Find perimeters of
polygons.

Finding Perimeters of Polygons

Jasmine wants to put a border around her vegetable garden. If she measures the
perimeter of the garden, Jasmine will know how much border to buy.

Terms to Know

Example / Illustration

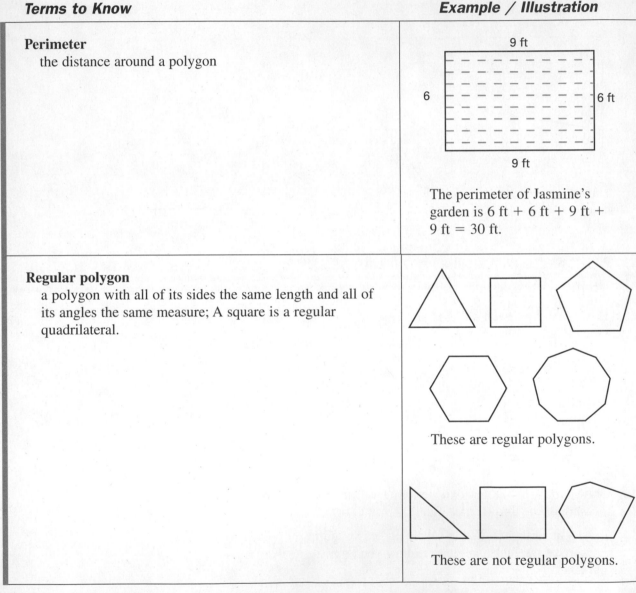

Perimeter
    the distance around a polygon

The perimeter of Jasmine's
garden is 6 ft + 6 ft + 9 ft +
9 ft = 30 ft.

Regular polygon
    a polygon with all of its sides the same length and all of
its angles the same measure; A square is a regular
quadrilateral.

These are regular polygons.

These are not regular polygons.

UNDERSTANDING THE MAIN IDEAS

You can find the perimeter of any polygon by finding the length of each side and
calculating the sum of the lengths.

(continued)

Lesson 3 *(continued)*

Example 1

Emilio draws plans for an enclosure for his dog on grid paper, as shown at the right. The edge of each grid square represents a length of 2 meters. How much fencing will Emilio need for the enclosure?

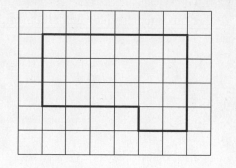

■ Solution ■

Step 1: Find the length of each side. Count by 2s for the edge of each grid square. The lengths are shown at the right.

Step 2: Add the lengths to find the perimeter.

$12 + 8 + 4 + 2 + 8 + 6 = 40$

Emilio will need 40 meters of fencing.

12 m

6 m   8 m

8 m   2 m

4 m

Find the perimeter of each polygon.

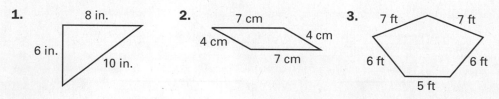

1. 8 in. 6 in. 10 in.

2. 7 cm 4 cm 4 cm 7 cm

3. 7 ft 7 ft 6 ft 6 ft 5 ft

Measure each side of the polygon to the nearest centimeter. Then find the perimeter.

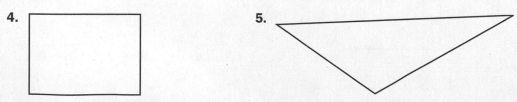

4.

5.

To find the perimeter of a regular polygon, multiply the length of one side by the number of sides.

(continued)

Example 2

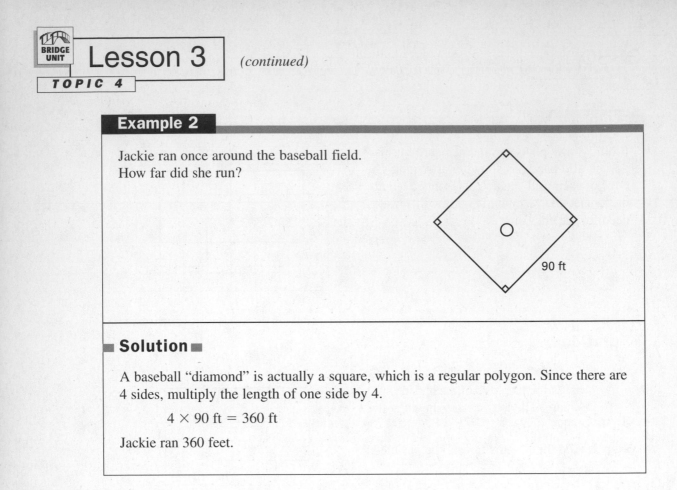

Jackie ran once around the baseball field.
How far did she run?

90 ft

▪ Solution ▪

A baseball "diamond" is actually a square, which is a regular polygon. Since there are 4 sides, multiply the length of one side by 4.

$$4 \times 90 \text{ ft} = 360 \text{ ft}$$

Jackie ran 360 feet.

Use multiplication to find the perimeter of each regular polygon.

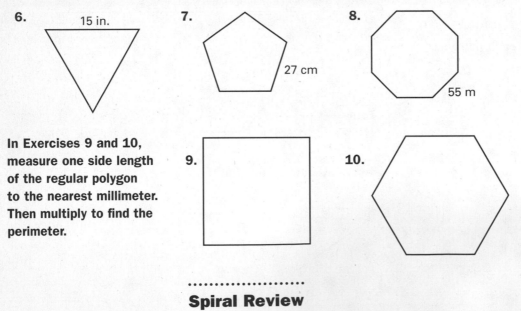

6. 15 in.

7. 27 cm

8. 55 m

In Exercises 9 and 10, measure one side length of the regular polygon to the nearest millimeter. Then multiply to find the perimeter.

9.

10.

Spiral Review

11. Write the fraction $\frac{40}{25}$ as a mixed number in simplest form and as a decimal.

12. What is the greatest common factor of 8, 10, and 12?

13. Kameisha has 48 tiles to arrange into a rectangular design. List the possible tile arrangements Kameisha can use.

Warm-Ups

Standardized Testing Warm-Ups

1. A regular nonagon (polygon with 9 sides) has one side with a length of 14 cm. What is the perimeter of the nonagon?

    A 136 cm   **B** 126 cm   **C** 196 cm   **D** not enough information

2. Which regular polygon below does *not* have a perimeter of 36?

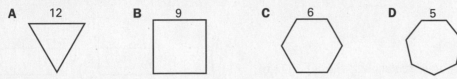

Homework Review Warm-Ups

3. Use a protractor to draw angles with measures of 18° and 142°.

4. What is each corner point of a polygon called? What are all the corner points together called?

Project

In Lesson 1 of Topic 4, you sketched the tables, shelves, and so on that you would be using for displaying goods and helping customers at the school store. To make the school store more attractive, you decide to put paper skirting around each rectangular table that you are using.

1. Using the measurements you made in Lesson 1, calculate the perimeters of all the tables you sketched.

2. Describe how you found the perimeter of each table. Did you add or multiply? Did you use paper and pencil or a calculator?

3. Suppose paper skirting comes in 100-foot bolts. How many bolts will you need in order to skirt every table in the school store?

BRIDGE UNIT

Activity 3

EXPLORATION

Cooperative Learning

Perimeter Patterns

In this activity, you will explore some of the patterns that will make the perimeters of rectangles easier to find and understand.

You will need:

- grid paper or dot paper

What to do:

Work with a partner.

1. a. On grid or dot paper, draw rectangles with a height of 1 and lengths of 1, 2, 3, 4, and 5, as begun at the right.

b. Find the perimeters of each of the rectangles you have drawn. Enter the numbers in the first "Perimeter" column of a table like the one below. For each size rectangle, the height is listed first.

c. Draw rectangles with a height of 2 and lengths of 1, 2, 3, 4, and 5. Find and enter their perimeters in the second "Perimeter" column of the table. Then do similarly for rectangles with heights of 3 and 4.

Size	Perimeter	Size	Perimeter	Size	Perimeter	Size	Perimeter
1 by 1		2 by 1		3 by 1		4 by 1	
1 by 2		2 by 2		3 by 2		4 by 2	
1 by 3		2 by 3		3 by 3		4 by 3	
1 by 4		2 by 4		3 by 4		4 by 4	
1 by 5		2 by 5		3 by 5		4 by 5	

2. a. What pattern do you see when the height stays the same but the length increases by 1? Is this true no matter what the height is?

b. Use your pattern to find the next row of the table.

c. Use your pattern to find the perimeter of rectangles with a height of 2 and lengths of 7, 8, and 9.

3. a. What pattern do you see when the height increases by 1 but the length stays the same? Is this true no matter what the length is?

b. Use your pattern to find the next perimeter column of the table.

c. Use your pattern to find the perimeter of rectangles with a length of 3 and heights of 6, 7, and 8.

4. Examine your drawings and their perimeters. In your own words, write a rule using multiplication for finding the perimeter of any rectangle given its height and length.

Basic Practice 3

FOR USE WITH TOPIC 4, LESSON 3

Find the perimeter of each polygon. If only one measurement is given, assume that the polygon is regular.

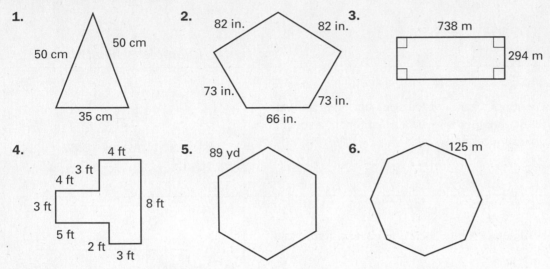

1. 50 cm 50 cm 35 cm

2. 82 in. 82 in. 73 in. 73 in. 66 in.

3. 738 m 294 m

4. 4 ft 3 ft 4 ft 3 ft 8 ft 5 ft 2 ft 3 ft

5. 89 yd

6. 125 m

Measure each side to the nearest millimeter and then find the perimeter. If you find that a polygon is regular, indicate this, and tell how you could find the perimeter more quickly.

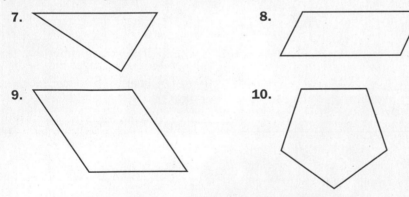

7.

8.

9.

10.

11. The Pentagon Building in Washington, D.C. has the shape of a regular pentagon. Each side of the building measures about 281 m. Suppose you could walk right around the outside of the building.

   a. Would you walk more than or less than a kilometer? Explain.

   b. How far would you walk?

Goal
Find the area of rectangles using square units.

Finding Areas of Rectangles

Jasmine is laying a new tile floor in her bathroom. The bathroom measures 8 feet by 6 feet. Jasmine needs to find the area of the bathroom floor so that she can buy the right amount of tiling.

Terms to Know	*Example / Illustration*
Square unit a measurement unit that is exactly one unit long and one unit wide; A square unit may be a square inch, a square meter, a square mile, and so on.	1 ft 1 ft ▨ Each tile measures 1 square foot, or 1 ft^2.
Area the number of square units needed to cover a figure in a plane	8 ft 8 ft It takes 48 tiles to cover the floor. The area is 48 square feet, or 48 ft^2.

UNDERSTANDING THE MAIN IDEAS

There are formulas for finding the area of many different polygons. For the bathroom floor, notice that the area is 48 square feet, and the length times the width is $6 \times 8 = 48$.

Area of a rectangle: Area = length $\times$ width, or $A = l \times w$

Remember that you always give an area in square units.

(continued)

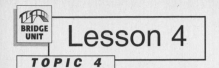
Example 1

Everyone in Mr. Greenley's family is piecing together squares to make a family patchwork quilt. The section Mr. Greenley is working on is shown at the right. How many squares must he piece together?

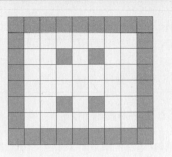

■ Solution ■

Step 1: You can think of each square as one square unit in the rectangular section. Find the length and width of the section.

The number of rows gives the length. The length is 8 units.

The number of squares in each row gives the width. The width is 9 units.

Step 2: Use the area formula for a rectangle.

$A = l \times w = 8$ units $\times$ 9 units $= 72$ square units

Because each square unit represents a quilt square, Mr. Greenley must piece together 72 squares.

Use the formula for the area of a rectangle to find the number of squares in each quilt piece.

1.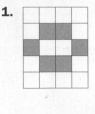

2.

3.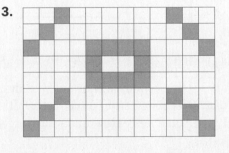

To apply the area formula, you don't actually have to show or picture the square units. All you need to know is the length and width of the rectangle.

(continued)

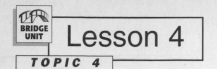

Example 2

Melika is buying carpet for her family room and bedroom. The rectangular family room measures 16 ft long by 24 ft wide. The bedroom is a square with each side of length 14 ft. How much carpet does she need?

■ Solution ■

Step 1: Find the area of the family room.

$$A = l \times w = 16 \text{ ft} \times 24 \text{ ft} = 384 \text{ ft}^2$$

Step 2: Find the area of the bedroom.

A square is a rectangle with all sides the same length. To find the area, you just have to multiply the length of a side times itself.

$$A = l \times w = s \times s = 14 \text{ ft} \times 14 \text{ ft} = 196 \text{ ft}^2$$

Step 3: Add the areas of the two rooms.

$$\text{Total area} = 384 \text{ ft}^2 + 196 \text{ ft}^2 = 580 \text{ ft}^2$$

Melika needs 580 square feet of carpet.

In Exercises 4–9, use the formula to find the area of each rectangle. Remember to record your answers in *square units*.

4.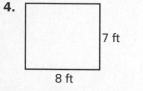
8 ft, 7 ft

5.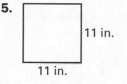
11 in., 11 in.

6.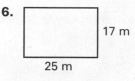
25 m, 17 m

7. square, side of 4 cm   **8.** length 19 m, width 12 m   **9.** square, side of 16 ft

10. Nguyen is roofing a house. The roof is in three pieces. Two pieces are rectangular, with lengths of 16 feet and widths of 46 feet. The other is a square 13 feet on a side. How many square feet of shingles does Nguyen need?

....
Spiral Review

11. Find a fraction equivalent to 0.64. Then write the fraction in simplest form.

12. Estimate the measure of the angle at the right to the nearest 5°.

(continued)

Warm-Ups

FOR USE WITH TOPIC 4, LESSON 4

Standardized Testing Warm-Ups

1. A rectangle measures 16 ft by 14 ft. What is its area?

   A 224 feet   **B** 60 feet   **C** 224 square feet   **D** 196 square feet

2. Which statement about the area of a polygon is *not* true?

   A The area of any polygon always represents some type of square units.

   B The area of any quadrilateral is found by multiplying its length times its width.

   C The area of a rectangle is the number of squares one unit on each side that it would take to cover the rectangle.

   D The area of a square is found by multiplying the length of a side by itself.

Homework Review Warm-Ups

3. A polygon's _____ can be found by computing the distance around it.

4. Counting the end zones, a football field is 360 feet long and 160 feet wide. How far would you walk if you walked around the outside edge?

Project

FOR USE WITH TOPIC 4, LESSON 4

In Lesson 1 of this topic, you sketched a layout of the school store. You included measurements for floor space and display areas.

1. Use the measurements you made of the classroom in Lesson 3 of Topic 1 to find the area of the school store to the nearest square meter.

2. Use your measurements from Lesson 1 of this topic to find the area of each rectangular table or display shelf. Which items should you display on the table(s) with the greatest area? the smallest area?

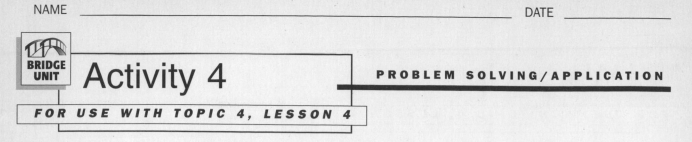

BRIDGE UNIT

Activity 4

FOR USE WITH TOPIC 4, LESSON 4

Areas of Irregular Polygons

In this activity, you will use what you know about the areas of rectangles and squares to find the area of an irregular polygon. If you can divide the polygon into smaller rectangular or square shapes, then you can find the area of the polygon by adding up the areas of the rectangles or squares.

Beth is buying carpet for her family room. Use the drawing of her family room to find how much carpeting she needs.

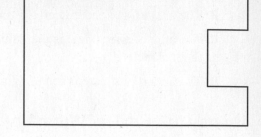

You will need:

- a centimeter/millimeter ruler
- plain paper

What to do:

1. Use a ruler to find the length of each side. Measure to the nearest millimeter.

2. Sketch the family room. Record your measurements on the sketch. Count each millimeter that you measure as 1 meter for the family room.

3. Think about how you can separate the area of the family room into smaller rectangular or square pieces. Show at least two different arrangements.

4. For each arrangement, calculate and record the areas of all the smaller rectangular and square pieces. Then calculate the total area by finding the sum of the smaller areas.

5. Compare your results. Are they the same?

6. You can also find the area by subtracting.

   a. Find the area of the large rectangle whose length and width are the longest length and width of the family room. Give the area in square meters.

   b. Find the area of the rectangular "cut-out" that is not carpeted.

   c. Subtract areas to find the amount of carpet needed. How does your result compare with the results you found by adding areas?

BRIDGE UNIT

Basic Practice 4

Use the formula for the area of a rectangle to find the number of tile squares in each design.

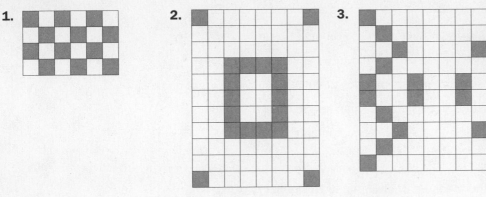

1.   **2.**   **3.**

In Exercises 4–9, use the formula to find the area of each rectangle.
Remember to record your answers in *square units*.

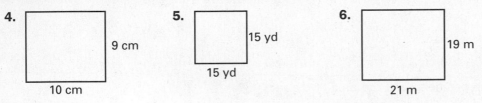

4. 9 cm, 10 cm   **5.** 15 yd, 15 yd   **6.** 19 m, 21 m

7. square, side of 13 m   **8.** square, side of 30 ft   **9.** length 13 yd, width 20 yd

10. A basketball court is 90 feet long and 54 feet wide. What is its area?

11. A tennis court is 78 feet long and 36 feet wide. What is its area?

12. A rectangle has an area of 24 square centimeters. List three possibilities for its length and width.

Measure each side to the nearest millimeter. Then find the total area.

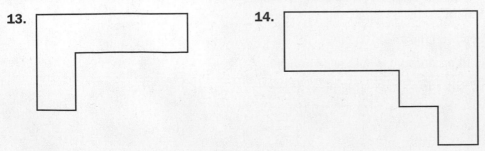

13.   **14.**

Basic Assessment

FOR USE WITH TOPIC 4

In Exercises 1–4, sketch the quadrilateral described. Then identify it using its most general name. Use the words in the box to help you.

square   rectangle   parallelogram   rhombus   trapezoid

1. two pairs of parallel sides, two pairs of sides of equal length, four right angles

2. two pairs of parallel sides, four sides of equal length, four right angles

3. two pairs of parallel sides, four sides of equal length

4. exactly one pair of parallel sides

Use a protractor to find the measure of each angle to the nearest 5°.

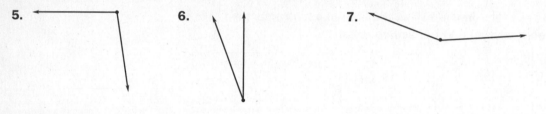

5.   **6.**   **7.**

In Exercises 8–10, find the perimeter and the area of the rectangle.

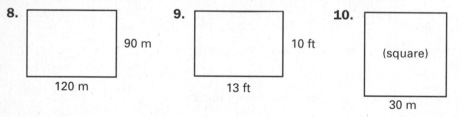

8.   90 m   **9.**   10 ft   **10.**   (square)

120 m   13 ft   30 m

11. Mr. Glass is buying carpet for his office. He also buy wood molding to go around the edge of the room. His office is rectangular in shape, with a length of 14 ft and a width of 11 ft.

   a. How much carpet does Mr. Glass need to buy?

   b. How many feet of molding does Mr. Glass need to buy?

12. A regular polygon with 7 sides has a side with a length of 9 cm. What is the perimeter of the polygon?

Answers

Lesson 1

Lesson Exercises *pages 106–108*

1. Yes.   2. Yes; a rectangle.

3. No; it has only one pair of parallel sides.

4. Yes; a square.

5. No; it does not have four sides of equal length.

6. No; it does not have four sides of equal length.

7. Yes; a square.

8. Yes.

9. A square has four sides of equal length; a rectangle has two pairs of sides of equal length.

10. A parallelogram has two pairs of parallel sides; a rhombus is a parallelogram, but its sides must also all be the same length.

11. Both have two pairs of parallel sides and four sides of equal length.

12. A trapezoid is the only special quadrilateral with exactly one pair of parallel sides.

13. **a.** trapezoid   **b.** parallelogram, rectangle
    c. parallelogram, rhombus, rectangle, square

14. about 10,000   15. 2, 3, 5, 9, 10

Warm-Ups *page 109*

1. D   2. C

3. $1\frac{1}{4}$; 1.25   4. $1\frac{1}{10}$; 1.1

5. $1\frac{7}{20}$; 1.35   6. $3\frac{1}{25}$; 3.04

Project *page 109*

Check students' work.

Activity *page 110*

1. Check students' work.

2. Sample answers are given.

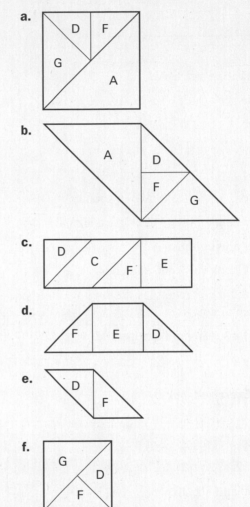

3. Students' designs will vary.

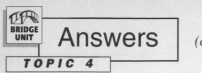

4. Sample answer:

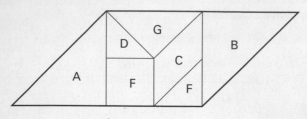

Basic Practice *page 111*

1. True.   **2.** False.

3. True.   **4.** True.

5. False.   **6.** False.

7. True.   **8.** False.

9. True.   **10.** a square

11. a rectangle   **12.** a trapezoid

13. a parallelogram

14–17. Check students' sketches.

18. Joan. Some rectangles do not have four sides of equal length.

19. two; a square and a trapezoid

Lesson 2

Lesson Exercises *page 114*

1. 70°   **2.** 150°   **3.** 25°

4–7. Check students' angles.

8. $2 \times 2 \times 2 \times 2 \times 2 \times 2 \times 2$

9. 2700 and 30; 90

Warm-Ups *page 115*

1. C   **2.** B

3. A rhombus and a square both have four sides of equal length and two pairs of parallel sides. A square also has four right angles.

4. trapezoid; no

Project *page 115*

1. Answers will vary.

2. a. Possible answers: hours of operation, goods to be sold, selling costs of items, location of store

b. Possible answers: the store's cost and predicted profit for each item

Activity *page 116*

1–3. Check students' work. Students may draw four out of five of the following figures: rectangle, square, parallelogram, rhombus, trapezoid. The sum of the measures of the six angles in each quadrilateral should be 360°.

4. a. 180°; The sum of the measures of the angles in a triangle is 180°.

b. The sum of the angle measures in each triangle is one half of the sum of the angle measures of the quadrilateral; This makes sense because the angles in the quadrilateral include all the angles in both triangles, and 180° + 180° = 360°.

c. It would be 360°. The diagonal just divides two of the angles of the quadrilateral each into two angles. Each pair of new angles exactly fits one of the old angles.

Basic Practice *page 117*

1. 85°   **2.** 50°   **3.** 150°

4. 90°   **5.** 35°   **6.** 115°

7–10. Check students' drawings and measurements.

11. $\angle 1 = \angle 3 = \angle 4 = \angle 6 = 45°$; $\angle 2 = \angle 5 = 90°$; sum = 360°

12. $\angle 1 = \angle 3 = 135°$; $\angle 2 = \angle 4 = 45°$; sum = 360°

13. 120°; Possible answer: 180° − (30° + 30°) = 120°; (or, students may sketch and measure).

14. 90°; Possible answer: 90° + 90° + 90° + 90° = 360°; (or, students may sketch and measure).

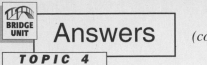
Lesson 3

Lesson Exercises *pages 119–120*

1. 24 in.   2. 22 cm   3. 31 ft

4. sides: 2 cm, 2 cm, 3 cm, 3 cm; perimeter: 10 cm

5. sides: 3 cm, 4 cm, 6 cm; perimeter: 13 cm

6. 45 in.   7. 135 cm   8. 440 m

9. 28 mm on each side; 112 mm

10. 18 mm on each side; 108 mm

11. $1\frac{3}{5}$; 1.6   12. 2

13. 6 by 8, 4 by 12, 3 by 16, 2 by 24, 1 by 48

Activity *page 122*

1.

Size	Perimeter	Size	Perimeter	Size	Perimeter	Size	Perimeter
1 by 1	4	2 by 1	6	3 by 1	8	4 by 1	10
1 by 2	6	2 by 2	8	3 by 2	10	4 by 2	12
1 by 3	8	2 by 3	10	3 by 3	12	4 by 3	14
1 by 4	10	2 by 4	12	3 by 4	14	4 by 4	16
1 by 5	12	2 by 5	14	3 by 5	16	4 by 5	18

2. **a.** The perimeter increases by 2; yes.

   b. The perimeters are 14, 16, 18, 20.

   c. 18, 20, 22

3. **a.** The perimeter increases by 2; yes.

   b. 12, 14, 16, 18, 20

   c. 18, 20, 22

4. Sample answers: To find the perimeter, multiply each dimension by 2. Then add the products.; To find the perimeter, add the length and height. Then multiply this sum by 2.

Warm-Ups *page 121*

1. B   2. D

3. Check students' drawings and measurements.

4. a vertex; the vertices

Project *page 121*

1. Answers will vary, but should reflect measurements from Lesson 1.

2. Calculation methods will vary.

3. Answers will vary, but should reflect an estimate of the quotient of the total sum of all table perimeters divided by 100, and then rounded up if necessary.

Basic Practice *page 123*

1. 135 cm   2. 376 in.   3. 2064 m

4. 32 ft   5. 534 yd   6. 1000 m

7. sides: 33 mm, 28 mm, 17 mm; perimeter: 78 mm

8. sides: 32 mm, 32 mm, 13 mm, 13 mm; perimeter: 90 mm

9. sides: each 26 mm; perimeter: 104 mm

10. sides: each 17 mm; perimeter: 85 mm; regular 5-sided polygon (pentagon); You can find the perimeter by multiplying the length of a side by 5.

11. **a.** More than a kilometer; 1 km = 1000 m and 281 × 5 > 1000.

    b. 1405 m, or 1.405 km

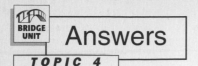

Lesson 4

Lesson Exercises *pages 125–126*

1. 20   **2.** 70   **3.** 96

4. 56 ft^2 **5.** 121 in.^{2} **6.** 425 m^2

7. 16 cm^2 **8.** 228 m^2 **9.** 256 ft^2

10. 1641 ft^2 **11.** $\frac{64}{100}$, $\frac{16}{25}$

12. 170°

Warm-Ups *page 127*

1. C   **2.** B

3. perimeter   **4.** 1040 ft

Project *page 127*

1. Answers will vary, but should reflect the product of the length and the width of the classroom measured in Topic 1, Lesson 3.

2. Answers will vary, but should reflect the products of the lengths and widths of the tables or shelves measured in Topic 4, Lesson 1.

Activity *page 128*

1. going clockwise from the top: 58 mm, 10 mm, 10 mm, 15 mm, 10 mm, 10 mm, 58 mm, 35 mm

2. Use the same numbers as in Step 1, but replace "mm" with "m."

3. Possible arrangements:

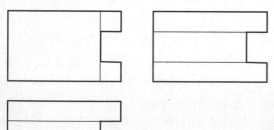

4. Areas of smaller figures will vary.; total area = 1880 m^2

5. The results should be the same.

6. a. 2030 m^2

    b. 150 m^2

    c. 1880 m^2; The results are the same.

Basic Practice *page 129*

1. 24   **2.** 88   **3.** 100

4. 90 cm^2 **5.** 225 yd^2 **6.** 399 m^2

7. 169 m^2 **8.** 900 ft^2 **9.** 260 yd^2

10. 4860 ft^2 **11.** 2808 ft^2

12. Possible answers: 6 cm by 4 cm, 8 cm by 3 cm, 12 cm by 2 cm, 24 cm by 1 cm

13. Total area: 550 mm^2

14. Total area: 1050 mm^2

Topic 4 Basic Assessment *page 130*

1–4. Check students sketches.

1. rectangle   **2.** square

3. rhombus   **4.** trapezoid

5. 100°   **6.** 20°   **7.** 155°

8. perimeter: 420 m; area: 10,800 m^2

9. perimeter: 46 ft; area: 130 ft^2

10. perimeter: 120 m; area: 900 m^2

11. a. 154 ft^2 **b.** 50 ft

12. 63 cm

Project Summary

FOR USE WITH BRIDGE UNIT

Now that you have done your research and made your decisions, you are ready to present your school store plan and advertisement to the class. Prepare a written or oral presentation involving all group members. Provide data from your research and measurements to support you decisions and conclusions. Be prepared to share your school store layout. Your report should include a summary of what you have learned while working on this project.

To understand how your work will be scored, consider the following scale from 1 to 4 (highest) points.

4 You have chosen items and used outside sources to determine their costs to you and the consumer. You have selected prices that are reasonable, and that offer a profit for the school store. Your scheduling reflects work hours and covers school store hours of operation. You have designed the store including carpeting, table skirting, and general floor layout. You have used your measurements to decide on the placement of each item to be sold. You have created a bulletin board advertisement for your store including all information that would be important for your customers to know. Your conclusions and decisions are wholly supported by your data. Your report is clear, understandable, and shows full understanding of the content learned in each topic.

3 You have chosen items and used outside sources to determine their cost to you and the consumer, but their pricing is either slightly too high to hold customers' interest or too low to yield a profit. Your scheduling does not reflect appropriate work hours or does not fully cover the entire time the school store is open. You have designed and decorated the store, but the plans are lacking in detail or in understanding of the measurements taken. The decisions you have make are somewhat supported by your data. Your report conveys your ideas well and shows some understanding of the content learned in each topic.

2 You have chosen items, but have not used outside sources to determine their cost. Your scheduling is vague or difficult to understand. The layout of your store and the measurements are not as exact as they should be. Your decisions appear reasonable, but you cannot fully support them with your data or measurements. Your report is incomplete.

1 Your plan is incomplete. You have not collected sufficient information on which to base your decisions. The school store layout is sketchy, lacks any mention of measurements, or is totally nonexistent. Projections for costs, sales prices, and profits are lacking or unreasonable. You have not prepared a report for the class. You should speak with your teacher as soon as possible to review your work and to make a new start on the project.

(continued)

Cumulative Assessment

FOR USE WITH BRIDGE UNIT

Round each number to the place held by its first digit to estimate each sum, difference, or product. Use compatible numbers to estimate each quotient.

1. $7231 + 8902$ **2.** $4.2 - 2.8$ **3.** 42×87

4. 19×28 **5.** $807 \div 89$ **6.** $2345 \div 61$

7. Rochelle drinks 520 milliliters of milk each day (there are 1000 milliliters in one liter). How many liter bottles of milk will Rochelle need to buy for one week's worth of milk?

Insert a digit so that the number is divisible by the given digits.

8. 41 _?_ ; divisible by 2 and 9   **9.** 7 _?_ 5; divisible by 5 and 9

Find the greatest common factor of each pair of numbers.

10. 6 and 22   **11.** 12 and 42   **12.** 16 and 40

13. What is the prime factorization of 100?

For each decimal, write a fraction or mixed number in simplest form.

14. 0.9   **15.** 0.44   **16.** 3.25

17. You are at a movie theater, watching a 135-minute film. Write a mixed number to describe the movie's length.

Draw an angle with the given measure.

18. 30°   **19.** 45°   **20.** 110°

21. A triangle has angle measures of 38° and 52°. What is the measure of the third angle?

22. What is the difference between a parallelogram and a trapezoid?

Find the perimeter and the area of each polygon.

23.   **24.**   **25.**

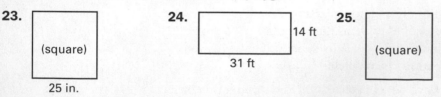

(square) — 25 in.

14 ft — 31 ft

(square)

Bridge Unit, PASSPORT TO MATHEMATICS BOOK 1

1. 16,000
2. 1
3. 3600
4. 600
5. 9
6. 40
7. 4 bottles
8. 4
9. 6
10. 2
11. 6
12. 8
13. $2 \times 2 \times 5 \times 5$
14. $\frac{9}{10}$
15. $\frac{11}{25}$
16. $3\frac{1}{4}$
17. $2\frac{1}{4}$ hr
18–20. Check students' work.
21. 90°
22. A parallelogram is a quadrilateral with two pairs of parallel sides. A trapezoid is a quadrilateral with only one pair of parallel sides.
23. perimeter: 100 in; area: 625 in.^{2}
24. perimeter: 90 ft; area: 434 ft^2
25. perimeter: 250 m; area: 3750 m^2